New Windows into the Universe

Arnold Hanslmeier

New Windows into the Universe

From the Hubble-, James Webb-, and Other Large Telescopes to Gravitational Wave Detectors

Arnold Hanslmeier
Institut für Physik Universität Graz
Graz, Österreich

ISBN 978-3-662-71371-6   ISBN 978-3-662-71372-3 (eBook)
https://doi.org/10.1007/978-3-662-71372-3

Die Deutsche Nationalbibliothek verzeichnet diese Publikation in der Deutschen Nationalbibliografie; detaillierte bibliografische Daten sind im Internet über https://portal.dnb.de abrufbar.

This book is a translation of the original German edition "Neue Fenster in das Universum" by Arnold Hanslmeier, published by Springer-Verlag GmbH, DE in 2023. The translation was done with the help of an artificial intelligence machine translation tool. A subsequent human revision was done primarily in terms of content, so that the book will read stylistically differently from a conventional translation. Springer Nature works continuously to further the development of tools for the production of books and on the related technologies to support the authors.

Translation from the German language edition: "Neue Fenster in das Universum" by Arnold Hanslmeier, © Der/die Herausgeber bzw. der/die Autor(en), exklusiv lizenziert an Springer-Verlag GmbH, DE, ein Teil von Springer Nature 2023. Published by Springer Berlin Heidelberg. All Rights Reserved.

© The Editor(s) (if applicable) and The Author(s), under exclusive license to Springer-Verlag GmbH, DE, part of Springer Nature 2025

This work is subject to copyright. All rights are solely and exclusively licensed by the Publisher, whether the whole or part of the material is concerned, specifically the rights of translation, reprinting, reuse of illustrations, recitation, broadcasting, reproduction on microfilms or in any other physical way, and transmission or information storage and retrieval, electronic adaptation, computer software, or by similar or dissimilar methodology now known or hereafter developed.
The use of general descriptive names, registered names, trademarks, service marks, etc. in this publication does not imply, even in the absence of a specific statement, that such names are exempt from the relevant protective laws and regulations and therefore free for general use.
The publisher, the authors and the editors are safe to assume that the advice and information in this book are believed to be true and accurate at the date of publication. Neither the publisher nor the authors or the editors give a warranty, expressed or implied, with respect to the material contained herein or for any errors or omissions that may have been made. The publisher remains neutral with regard to jurisdictional claims in published maps and institutional affiliations.

This Springer imprint is published by the registered company Springer-Verlag GmbH, DE, part of Springer Nature.
The registered company address is: Heidelberger Platz 3, 14197 Berlin, Germany

If disposing of this product, please recycle the paper.

Preface

Astronomy is the oldest of the natural sciences. The study of the stars was already carried out in ancient cultures for a very practical need: a division of time, a *calendar,* was needed, and for seafaring or extensive travels on land, *navigation* was essential. Both are based on astronomical processes such as the orbit of the Earth around the Sun (one year), the orbit of the Moon around the Earth (about one month), an Earth rotation (one day), or the height of the celestial pole (Polaris, currently indicates the geographical latitude of the observation site). In addition to this classical astronomy, *astrophysics* has developed particularly since the invention of the telescope, the discovery of the law of gravitation and other laws of celestial mechanics, as well as the explanation of radiation and spectral lines.

In the first parts of the book, these fundamental laws are explained with simple means. For particularly interested readers, there are always short mathematical interludes, which are specially marked, but can be omitted - without losing the context. Modern large observatories are introduced: the telescopes of the European Southern Observatory ESO, the telescopes in Hawaii and on the Canary Islands; the 40-m telescope of the ESO, which is still under construction, is also presented. With these telescopes, groundbreaking discoveries are still being made and achieved: exoplanets around other stars have been discovered, in rare cases even directly photographed. The curvature of space-time predicted by the *general theory of relativity* could be directly observed using distorted extremely distant galaxies. Stars near the center of our Milky Way showed a movement within a few years around something we cannot see: a *supermassive black hole.*

Looking into the depths of the universe is always also a look into the past due to the finite speed of light. Just over 100 years ago, it was thought that the universe was as large as the Milky Way, but E. Hubble was able to show with a telescope that is only medium-sized by today's standards, that there are billions of galaxies, similar to the Milky Way consisting of several 100 billion stars, and that the universe is expanding.

But what we see with our eyes is only a tiny fraction of the entire electromagnetic spectrum. We first address the question of what light actually is and encounter quantum physics in the process. We explain the dark lines in the dispersed light, which allow us to specify the composition of galaxies that are billions of light years away from us. And we show from which time in the development of the universe the first light comes that we can receive; due to the strong redshift, this radiation can be observed in the microwave range. We see here the first structures in the universe.

New windows into the universe means that today we can study the cosmos in practically all wavelengths: from the extremely short-wave X-rays to the long radio waves. We have even achieved what Einstein, the founder of the general theory of relativity, did not think possible: gravitational waves have been directly detected; these were created by the collision of two black holes several 100 billion light years away from us. Since the Earth's atmosphere only allows visible light or a small range in the infrared and part of the radio range, we rely on telescopes in space. The first large space telescope was the *Hubble Telescope,* which was launched into Earth orbit by a space shuttle, which turned out to be a great fortune, as it had to be repaired to really get the desired sharp images. Currently, the *James Webb Telescope, JWST* is delivering amazing images, predominantly in the infrared range. We will show why this wave range is so significant for astrophysicists. But there are also huge astronomical projects on Earth. Exotic-looking detectors, with the help of which we can detect neutrinos from the sun and supernovae more than 100,000 light years away. Globally networked radio telescopes, with which we received a "picture" of the supermassive black hole located in the center of galaxies for the first time, etc.

One thing is beyond question: the more sensitive the instruments, the more details will be revealed, and perhaps even the three big questions of modern astrophysics will be solved:

- Does dark matter exist and how can it be explained?
- Does dark energy exist and how can it be explained?
- Are we alone in the universe?

Modern astrophysics is unthinkable without the insights of relativity theory and quantum physics. However, it has also significantly contributed to the further development of modern physics. I invite you, dear reader, on a foray into modern astrophysics, modern physics, which offers us new windows into the universe, allows us to discover fantastic things, leads to astonishment, but also raises many new questions. An introduction to modern astrophysics can be found in [5] or [9] and [13].

Again, it should be emphasized: one can read the book chapter by chapter, skip the mathematical inserts and examples, or of course also take pleasure in them.

I would like to express my sincere thanks to the editorial team of Springer Publishing for their excellent collaboration, and to all my colleagues who have accompanied me in numerous discussions. Special thanks also to my partner Anita, my son Roland who has accompanied me musically at some public lectures, as well as to my daughters Christina and Alina, their families, and to Jacqueline.

Graz, Bad Gleichenberg　　　　　　　　　　　　　　　　　　　　　　　Arnold Hanslmeier
May 2023

Contents

1 The Light of the Stars 1
    1.1 Sun—Stars—Planets 1
        1.1.1 The Sun 2
        1.1.2 Planets 4
        1.1.3 Stars 4
    1.2 The Radiation of Stars and Planets 5
        1.2.1 A Law for Radiation? 5
        1.2.2 The Brightness of the Stars is Measured 6
        1.2.3 Distances of the Stars 7
        1.2.4 The Radiation of Planets 8
    1.3 What is Light Actually 9
        1.3.1 Light as a Particle 9
        1.3.2 Light as a Wave 11
        1.3.3 Light and Quantum Physics 13
        1.3.4 Spectral Lines in the Spectrum 15
        1.3.5 The Doppler Effect 16
        1.3.6 Measuring Speeds in the Universe 17

2 Telescopes—Eyes into the Universe 21
    2.1 Basic Types of Telescopes 21
        2.1.1 Lenses 21
        2.1.2 Mirrors 25
        2.1.3 Lens and Mirror Telescopes 26
        2.1.4 Magnification and Resolution of a Telescope 27
    2.2 The Earth's Atmosphere 30
        2.2.1 The Permeability of the Earth's Atmosphere 30
        2.2.2 When do the Sun and Moon Rise? 31
        2.2.3 The Turbulent Earth's Atmosphere or Why Stars Twinkle 34
        2.2.4 The Human Eye: Our Natural Detector 34
        2.2.5 CCD 37

		2.2.6	Photography	38
		2.2.7	The View into the Telescope	38
	2.3	The Brightness of Stars		39
		2.3.1	How Bright Do Stars Actually Shine	39
		2.3.2	The Absolute Brightness	40
	2.4	Some Examples of Historical Telescopes		42
		2.4.1	From the Invention of the Telescope to the First Large Telescopes	42
		2.4.2	Internship for Enthusiasts: Observing Sunspots	45
		2.4.3	New Planets in the Solar System	47
	2.5	Modern Large Telescopes		50
		2.5.1	The Focal Length Determines the Image Size	50
		2.5.2	The Largest Solar Telescope	50
		2.5.3	The European Southern Observatory, ESO	52
		2.5.4	Observatories on the Canary Islands	54
		2.5.5	Telescopes in Hawaii	60
3	**Light from the Edge of the Universe—The World of Galaxies**			65
	3.1	Size and Expansion of the Universe		65
		3.1.1	Our Milky Way—Home in the Cosmos	65
		3.1.2	Galaxies—Building Blocks of the Universe	67
		3.1.3	The Expansion of the Universe	72
		3.1.4	The Universe Had a Beginning	74
	3.2	The Universe is Getting Colder		75
		3.2.1	Experiment With the Bicycle Pump	75
		3.2.2	Looking Into the Past	76
		3.2.3	Glow From the Time of the Big Bang	78
		3.2.4	Tracing the First Light	81
	3.3	The Origin of Chemical Elements		83
		3.3.1	The First Three Minutes—or the Primordial Nuclear Fusion	83
4	**The Radio Sky**			87
	4.1	What is Observed in the Radio Range		87
		4.1.1	Radar Signals Explore Planets	87
		4.1.2	Discovery of Radio Emission from the Universe	91
		4.1.3	How a Radio Telescope Works	91
		4.1.4	How Radio Emission Occurs	93
	4.2	Black Holes		95
		4.2.1	Black Holes and Star Development	95
		4.2.2	Supermassive Black Holes	101
		4.2.3	Making the Monster Visible	102
		4.2.4	ALMA	104

5 Gravitational Waves—A New Window into the Cosmos 109
5.1 General Theory of Relativity 109
5.1.1 Einstein and Relativity 109
5.1.2 From Special to General Relativity 112
5.2 What are Gravitational Waves? 115
5.2.1 Waves 115
5.2.2 How Gravitational Waves Are Created 116
5.2.3 When Stars Collide 118
5.2.4 Collision of Two Neutron Stars 119
5.2.5 Properties and Detection of Gravitational Waves 120
5.2.6 The Spectrum of Gravitational Waves 121
5.3 The First Direct Observation of Gravitational Waves 122
5.3.1 Further Measurements 125
5.3.2 Dark Matter and Gravitational Waves 125

6 Neutrinos—Ghost Particles 129
6.1 What are Neutrinos 129
6.1.1 A Whole Zoo of Particles 129
6.1.2 How elementary are protons and neutrons 131
6.1.3 Neutrinos 135
6.2 Discovering Neutrinos 137
6.2.1 The First Neutrino Observatory 137
6.2.2 Additional Neutrino Telescopes 138
6.3 Where do Neutrinos from the Universe Come From 139
6.3.1 The Problem of Solar Neutrinos 139
6.3.2 Neutrinos from a Supernova Explosion 143

7 The Hubble Space Telescope 147
7.1 Planning, Launch, Problems 147
7.1.1 First Ideas 147
7.1.2 Precursor Missions 149
7.2 The Great Disappointment 151
7.2.1 An Error in Optics 151
7.2.2 The Error is Corrected 152
7.3 Instruments of the Hubble Telescope 154
7.3.1 What Does a Space Telescope Look Like? 154
7.3.2 Instruments on Board the HST 156
7.4 Some Observation Results With the Hubble Telescope 159
7.4.1 Solar System 159
7.4.2 Stars, Nebulae 160
7.4.3 Galaxies 166
7.4.4 A Look into the Early Universe: The Hubble Deep Field 168

8 The James Webb Telescope 175
8.1 Planning 175
8.1.1 First Ideas for the Successor to the Hubble Telescope 175
8.1.2 Difficulties 176
8.1.3 The Launch 177
8.2 Structure and Instruments of the JWST 178
8.2.1 Construction of the Telescope 178
8.2.2 Where is the James Webb Telescope located? 179
8.3 Basic Supply and Instruments 181
8.3.1 Supply Unit, Power Supply 181
8.3.2 Attitude Control, Communication System 182
8.3.3 Propulsion 183
8.3.4 The Sunshield 184
8.3.5 The Optics 184
8.4 Instruments of the JWST 185
8.4.1 The Near Infrared Camera 185
8.4.2 MIRI 186
8.4.3 NIRSpec, FGS-NIRIS 186
8.5 First Images 186
8.5.1 Solar System 187
8.5.2 Exoplanets 188
8.5.3 Stars 193
8.5.4 Galaxies and the Early Universe 195

References 203

The Light of the Stars   1

Astronomy distinguishes itself from other sciences by one peculiarity: we are passive observers. Due to the vast distances, planets, stars, galaxies can only be studied passively. The only source of information is their radiation. Therefore, from the radiation of celestial objects, one must deduce their physics and determine typical quantities such as:

- Distance,
- Mass,
- Temperature,
- Composition,
- Magnetic fields,
- Age, etc.

determine.

In this chapter, we want to explain some physical principles on how this information is obtained from the light of stars, etc.

1.1 Sun—Stars—Planets

In this section, we will discuss the sun, the stars, and the planets. What is the difference between a star and a planet? How can we get information about distant stars that we will probably never reach? Why do stars shine at all? In addition to these basic questions, it will also be shown how we get information from the radiation of the stars.

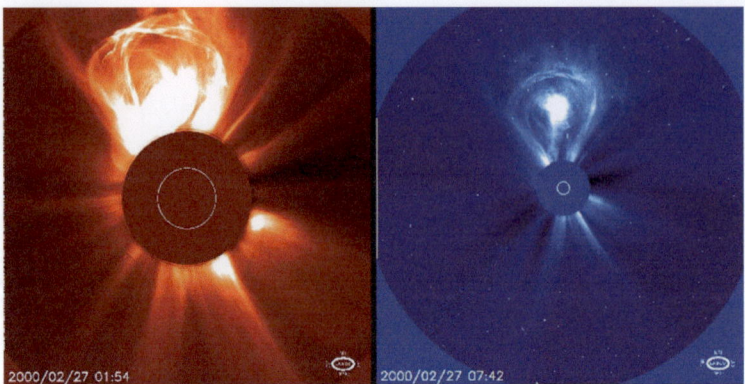

Fig. 1.1 The sun captured by a satellite. The image shows a coronal mass ejection. The solar disk itself is covered and sketched by the circle. Note the dimensions. The solar disk has 109 times the diameter of the earth. SOHO/ESA/NASA

1.1.1 The Sun

A popular question is: "what is the name of our nearest star". Then there are the most varied answers, e.g. Alpha Centauri or Proxima Centauri. Most, however, forget that the nearest star is our sun. The sun is the only star whose surface we can study in detail: we observe dark sunspots, whose diameter can exceed the diameter of the earth, huge matter eruptions, so-called *coronal mass ejections,* where billions of tons of matter are hurled into space away from the sun (Fig. 1.1), energy eruptions, so-called *flares,* in which energy is released within a few minutes that corresponds to several million hydrogen bomb explosions on earth and much more. The great importance of the sun, the star in front of our doorstep, for life on earth was already recognized by the ancient cultures and the sun was often worshiped as a god (e.g. the Egyptian sun god Ra).

Without the sun, there would be no life on Earth. But the study of the sun plays a major role in all of astrophysics. Activity phenomena such as sunspots, magnetic field loops, etc. (Fig. 1.2) can only be indirectly detected in other stars. But one can also compare the sun with other stars, leading to the question of whether our sun is a special star or whether many stars are similar to our sun. This is also important when we deal with the topic of *life in the universe*: from this has developed a whole branch of science, *astrobiology,* . It turns out that life, as we know it from Earth, presupposes a planet that orbits a sun-like star. By a sun-like star, we mean an object with about the mass of the sun, the temperature of the sun, and the expansion of the sun. These parameters for the sun are summarized in Tab. 1.1. The sun is the brightest object in the sky, but there are stars that shine up to 10,000 times and more brightly than the sun. The sun only appears to us as the brightest object because of its proximity. The distance Earth-Sun is about

1.1 Sun—Stars—Planets

Fig. 1.2 The sun captured by a satellite. The image shows the sun in short-wave UV light. The loop-shaped structures are caused by the arrangement of the hot solar plasma by magnetic fields in the higher atmosphere of the sun. Solar Orbiter, ESA/NASA

Tab. 1.1 Important data for our sun

Parameter	Value	relative to the Earth
Size	1 400 000 km	109
Mass	2×10^{30} kg	330 000
Temperature	Surface: 6000 K, Center: 12 million K	Earth: globally about 15 °C

150,000,000 km[1] . Light propagates at a speed of 300,000 km/s. One can easily calculate how long the light from the sun takes to reach the Earth:

$$t = \frac{150\,000\,000\,\text{km}}{300\,000\,\text{km/s}} \sim 500\,\text{s} \tag{1.1}$$

---

[1] This is also referred to as *Astronomical Unit, AU,*.

So, the light takes about 8 minutes from the sun to the Earth. When we observe the sun at this moment, we see light that was sent to us 8 minutes ago.

1.1.2 Planets

The sun is the largest body in the solar system. If you combine the mass of all 8 major planets (Mercury, Venus, Earth, Mars, Jupiter, Saturn, Uranus, Neptune), as well as all other celestial bodies (dwarf planets, asteroids, etc.), this mass only amounts to about 0.2% of the mass of the sun. This is also outlined in Fig. 1.3. This figure shows a size comparison of the sun with the planets. The distances from the sun are not to scale. From the sun's perspective, Earth is the third planet. In terms of size, our home planet is larger than Mercury, Mars, and Venus, but smaller than Jupiter, Saturn, Uranus, and Neptune. The largest planet in the solar system is Jupiter, with about 10 times the diameter of Earth. The Earth's diameter is about 12,756 km.

1.1.3 Stars

Our nearest star, the sun, is just one of several hundred billion stars in the *Milky Way*, the *Galaxy*. But these stars are much further away from us than the sun and therefore only appear as points even in the world's largest telescopes. This brings us to the biggest problem in astrophysics: we cannot experiment with the objects, the only information

Fig. 1.3 Comparison of the sun, planets, and other bodies of the solar system

we get directly from the stars is their radiation. The task of astrophysics is to derive all physically relevant information from the radiation of the stars. Stars shine by themselves, the necessary energy usually comes from nuclear fusion taking place inside a star, where lighter elements are fused into heavier elements.

▶ The sun is our nearest star and one of several hundred billion stars in the Milky Way, the galaxy.

1.2 The Radiation of Stars and Planets

1.2.1 A Law for Radiation?

It's no easy task to derive the physics of stars from the *radiation* of stars. At the beginning of the 20th century, the physicist *Max Planck* (1848–1947) was able to describe the radiation of a *black body* with a simple formula. What is meant by the term black body? The definition was given in 1859 by *G. R. Kirchhoff*. A black body is an object that absorbs all radiation and does not reflect. One could imagine a black body as a cavity that only has a tiny opening. But why do we deal with such a model? Fortunately, the radiation of most objects (stars, planets, etc.) is very similar to that of a black body, and we can therefore apply the laws of radiation to them.

Light can be broken down using a prism or grating. This is known, for example, from the rainbow. Here, sunlight is broken down into colors by fine water droplets in the Earth's atmosphere. The colors mean nothing other than different wavelengths. Blue light differs from red light only in the wavelengthng:

- blue light: wavelength around 400 nm[2]
- red light: wavelength around 600 nm.

The physicist *W. Wien* (1864–1928) was able to show, using Planck's law, that the product of the wavelength at which an object shines brightest and its temperature is constant. Wien's law states:

$$T\lambda_{max} = \text{const} \tag{1.2}$$

From this it follows: the hotter an object, the more it radiates in the short-wave range. Thus, with the *naked eye,* we are able to estimate the *temperature of stars.* Just look at bright stars in the sky. You will notice that there are stars that shine rather red, stars that shine yellow, and even stars that shine white to slightly blue. We now know what this

---

[2]$1 \text{nm} = 1 \text{ Nanometer} = 10^{-9}$ m = ein billionths of a meter.

Fig. 1.4 The double star Albireo in the constellation Cygnus. Two stars of different temperature. The temperatures are about 14,000 K and 4,000 K

means: blue stars are much hotter than yellow ones, which in turn are hotter than red ones.

▶ The color of the stars is a measure of their temperature.

- Blue stars: Temperature about 40,000 K[3]
- Yellow stars: Temperature about 6000 K
- Red stars: Temperature about 3000 K.

A beautiful example of two closely spaced stars is given in Fig. 1.4. The stars have different colors and thus different temperatures.

1.2.2 The Brightness of the Stars is Measured

How brightly stars shine depends on two factors: (i) their actual luminosity (ii) their distance. Stars can appear very bright to us, even though they actually have only low luminosity, because they are very close to us. Since ancient times, the *apparent brightness* of stars has been divided into magnitude classes: stars that appear brightest are referred to as first magnitude stars, and stars that are just visible to the naked eye (but only in complete darkness) as sixth magnitude stars. This scale has been extended to also capture very bright objects, so it also goes into the negative: the full moon, for example, has an apparent brightness of −12 magnitudesmagnitude classes.

---

[3] K stands for Kelvin; the Kelvin scale starts at the lowest temperature $0\,\text{K} = -273{,}2\,°\text{C}$.

1.2 The Radiation of Stars and Planets

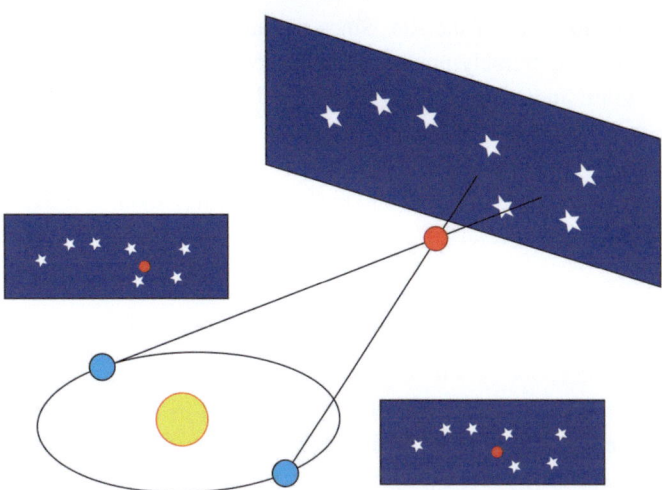

Fig. 1.5 The principle of annual parallax for determining the distance of nearby stars (red) that move against the sky background over the course of a year due to the Earth's movement around the sun

In addition to this apparent brightness, which is indicated by a superscript m^4, there is the *absolute brightness.* This refers to the apparent brightness that an object would have at a certain distance. The absolute brightness is indicated by a superscript M.

An important feature of *telescopes* is to collect as much light as possible. The larger the *aperture* (of the lens or mirror) of a telescope, the more light is collected and fainter objects can be detected.

1.2.3 Distances of the Stars

The distances of the stars have nothing to do with their properties. However, if one wants to study the structure of the galaxy, it is important to know the distances of the stars. A very simple method for determining the distance of stars is the *parallax method.* Parallax means, that the position of a nearby object observed from two separate points changes relative to a distant background.

Friedrich Wilhelm Bessel determined the parallax of the star *61 Cygni* in 1838. The Earth moves around the sun over the course of a year. Therefore, relatively nearby stars seem to move back and forth against a distant background over the course of a year (Fig. 1.5). The angle of the annual parallax is then given by the ratio of the radius of the

---

[4] m stands for magnitudo, lat. magnitude class.

Earth's orbit to the distance of the star. Since the angle is less than 1 arcsecond[5], it was only determinable by extremely precise measurements. With modern measurements carried out by satellites, angles under 1/1000" can be determined today.

The distance of the stars is given in light years. A light year is the distance that light travels in one year.

Light propagates at a speed of 300,000 km/s. A year has about 365 days, a day has 24 hours, an hour has 3600 seconds. Thus, a year has about 30 million seconds and we get for a light year:

$$1\,\text{Lj} = 3 \times 10^5 \text{km/s} \times 3 \times 10^7 \text{s} \sim 10^{13}\,\text{km} \qquad (1.3)$$

The star 61 Cygni is about 11.4 light years away from us. We remember: from the sun, light takes about 8 minutes to reach Earth, in the case of 61 Cygni, however, 11.4 years. A light year is the distance that light travels in one year. So if you want to know what happened on this star 11.4 years ago, you have to observe it tonight.

▶ A light year equals 10 trillion km = 10^{13} km

1.2.4 The Radiation of Planets

In contrast to stars, planets do not shine themselves or only very weakly. They are illuminated by the sun (solar system) or their parent star (exoplanets). Since ancient times, 5 planets are known: Mercury, Venus, Mars, Jupiter, and Saturn, as well as the Earth, whose position in the solar system was not known for a long time. There were two concepts:

- *geocentric worldview:* the Earth is at the center, everything moves around the Earth.
- *heliocentric worldview* (N. Copernicus 1543): the sun is at the center, Earth and planets move around it.

These 5 planets are visible to the naked eye. Uranus and Neptune were only discovered in the 18th and 19th centuries with telescopes. Planets that orbit other stars are always very close to the sky at their parent star due to the great distance to us and can hardly be seen. Their radiation is usually more in the infrared, as the temperatures are low. For this reason, the observation of exoplanets in this wavelength range is important. We remember Wien's law. The cooler a celestial body, the more the maximum of its radiation shifts into the red range and finally into the infrared range, which lies beyond red (Fig. 1.6).

---

[5] 1 arcsecond= 1" =1/3600 degrees.

Fig. 1.6 The Earth seen from space. Even from the next star (apart from the sun), Alpha Centauri, it would hardly be possible for a civilization with our current means to see the Earth directly

For mathematics/physics fans: let's calculate where the maximum of solar radiation (T = 6000 K) and the radiation of the Earth lies (T = 300 K). It applies (Wien's law)

$$T\lambda_{max} = \text{const} = 2897\,\mu m \qquad (1.4)$$

$1\mu m = 10^{-6}$ m. We find (rounded values): (i) Sun: $\lambda_{max} = 0{,}5\,\mu m = 500\,nm$, (ii) Earth: about $10\,\mu m$ thus in the infrared.

1.3 What is Light Actually

Light is electromagnetic radiation one could formulate somewhat dryly, but this hides great concepts of modern physics. For a long time, it was not clear what light actually was: a wave or a particle.

1.3.1 Light as a Particle

What happens during vision, what is light, how is it absorbed by the eye? There were so-called emission theories also known as *corpuscular theories*. It was imagined that light was nothing more than the propagation of some particles, which were referred to as corpuscles. The first known philosophers who dealt with light were *Pythagoras* (around 570 BC to about 510 BC) and *Empedocles* (ca. 495-435 BC). Further representatives of this theory were later *Augustine* (354-430, according to his view, rays in the form of particles go out from our eyes with which we scan the environment) and then the physicists/mathematicians *Isaac Newton* (1643–1727, Fig. 1.8), *Pierre Simon de Laplace* (1749–1827), and others.

Light has certain properties that a theory must explain: it propagates in a straight line, light rays can be reflected, etc. The corpuscular theory of light can explain these two phenomena. The different colors were explained by different sizes of the corpuscles. The diffraction of light was explained by a deflection of the light particles; these are attracted to obstacles.

Newton held the view that light particles would be attracted by masses. One should be able to measure a deflection of light by the gravitational field of the sun during a total solar eclipse. Stars that become visible near the sun darkened by the moon should show a slight shift in their position (Figs. 1.7 und 1.8).

The deflection of light by gravity was impressively confirmed 200 years later and also predicted by Einstein, however, as a result of space-time curvature. We will go into this later. Pierre Simon de Laplace went a step further and believed that there could be stars so massive that due to their strong attraction not even light could escape. We refer to such an object today as a *Black Hole*.

Johann Georg von Soldner (1776–1833) calculated the light deflection at the sun and found the value of 0.84. The value that results from Einstein's *general theory of relativity* is twice as large.

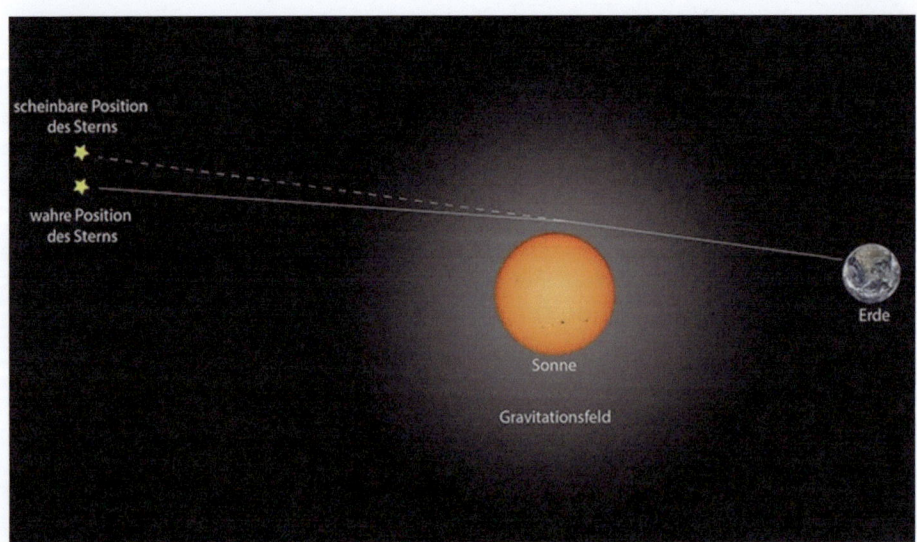

Fig. 1.7 The deflection of a light beam passing by the sun

1.3 What is Light Actually

Fig. 1.8 Isaac Newton, who believed that light could be explained by propagating particles

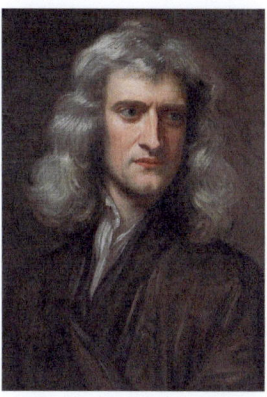

For those interested in mathematics: the deflection of a light beam, which passes at a minimal distance b by a mass M, according to the general theory of relativity is:

$$\Delta\phi = \frac{4GM}{bc^2} \quad (1.5)$$

where $G = 6{,}67 \times 10^{-11}$ is the gravitational constant and $c = 3 \times 10^8$ is the speed of light. This is outlined in Fig. 1.9.

1.3.2 Light as a Wave

Certain properties of light cannot be explained with the corpuscular theory, or only with the assumption of auxiliary hypotheses. This includes the *diffraction* of light as well as the *interference* of light.

Let's briefly explain how a wave can be described:

- Wavelength λ Distance between two wave peaks or wave troughs
- Displacement, *Amplitude, A*. The amplitude reaches the highest values at wave peak (positive) or wave trough (negative).

Fig. 1.9 Light deflection near a mass M

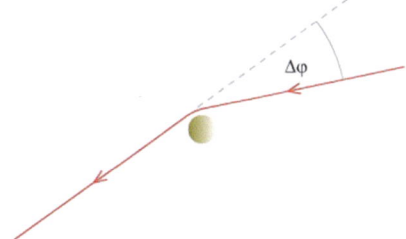

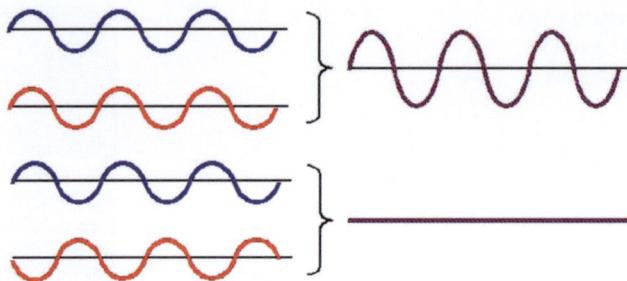

Fig. 1.10 Interference of two waves (red, blue). If the two waves are phase-shifted by 90 degrees against each other, their superposition results in the value zero

If we consider two waves that overlap, this is referred to as *interference* of waves. If the waves have the same wavelengths and the overlap occurs in such a way that wave peaks of the first wave always meet wave peaks of the second wave or the same with the wave troughs, then the resulting wave has a larger amplitude. In this case, both waves are referred to as waves with a phase shift of zero (Fig. 1.10 above).

But what happens if wave peaks of the first wave always coincide with wave troughs of the second wave? Then the resultant is zero, the two waves cancel each other out. The phase shift here is 90 degrees (Fig. 1.10 below).

The interference of two waves can be demonstrated, for example, with 2 laser beams or directly observed with water waves when, for example, two stones are thrown into calm water. If light is a wave phenomenon, then the interference of two light waves can be easily explained.

Another phenomenon of waves is their diffraction at obstacles. Let's do a simple thought experiment. If light were a stream of particles that, for example, goes through a window, how can it be that the entire room still becomes bright? Further diffraction phenomena occur when a wave goes through a gap. Let's consider the famous *double-slit experiment*. If light were a stream of particles, one would have to measure 2 accumulation points of the particles behind the slits. However, if light is a wave phenomenon, then due to diffraction a typical diffraction pattern occurs, which can be explained by the superposition of phase-shifted waves (Fig. 1.11 and 1.13).

▶ The idea that light consists of particles is supported by: reflection, propagation, impulse.

The idea that light is a wave is supported by: diffraction, interference, reflection.

1.3 What is Light Actually

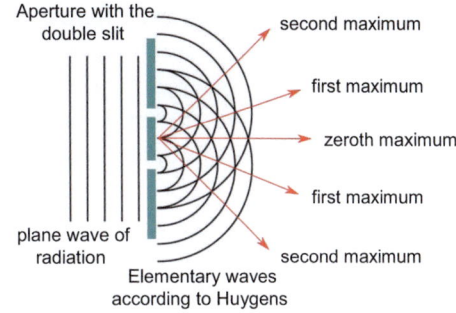

Fig. 1.11 A wave passes through a double slit. According to the *Huygens principle* every point is the starting point of an elementary wave, which explains the resulting pattern

1.3.3 Light and Quantum Physics

We now find ourselves in a strange situation. We wanted to explain whether light is a wave or a particle phenomenon. However, the result was that there are experiments that can only be interpreted with one of the two explanations.

We have already mentioned Planck's radiation laws, with the help of which one can calculate the radiation emitted by a star at any given wavelength. The emitted intensity depends only on the temperature of the object. But to find this law, Planck had to make a seemingly very strange assumption: he assumed that light consists of particles, and now comes the really new part: the energy of these particles only comes in portions, one says it comes in *quanta*. The energy of a light quantum, called a *photon*, is an integral multiple of the product of the frequency and a constant, which was later referred to as Planck's constant. Let's briefly consider what this means. The energy is not arbitrary, but always a multiple of a "basic energy".

The energy of a photon is given by

$$E = nh\nu \qquad (1.6)$$

$n = 1,2,3,4....$, ν the frequency of the light and $h = 6{,}626 \times 10^{-34}$ Js. The frequency is the number of oscillations per second and it applies

$$c = \lambda \nu \qquad (1.7)$$

In our familiar environment, this seems very strange: a particle can only have the energy $h\nu, 2h\nu, \ldots$ and no values are possible in between! Why don't we notice the quantization of energy? This is due to the smallness of the *Planck's quantum of action*. We usually cannot distinguish such small amounts of energy, therefore energy appears to us as continuous and not discrete.

Quantum physics goes even further; not only light particles, photons, possess this property but all quantum mechanical particles and there is a wave-particle duality. There are certain experiments where the particle character is more noticeable, and other experiments where the wave character comes into play. Photons, electrons, protons behave depending on the experiment like a wave or a particle, they are both, wave and particle.

The particle character of light can also be determined by other experiments. In the *photoelectric effect*[6] , light particles hit a metal surface, and knock electrons out of the atoms on the surface. This can be measured as current. But a certain energy threshold must be overcome in order for the electrons initially bound to the atoms to be released. Another effect is the *Compton effect*. A photon hits an electron, similar to two billiard balls colliding. The photon gives part of its energy to the electron. We already know the energy of a photon: it depends on its wavelength, the smaller the wavelength, the greater the energy. Short-wavelength light has higher energy than long-wavelength.

$$E \sim \frac{1}{\lambda} \tag{1.8}$$

If, therefore, the photon has a lower energy after the collision, its wavelength must have increased. This can be measured exactly. The next step in quantum physics was then relatively small: *de Broglie* (1892–1987) postulated that all particles possess wave properties.

> The *de Broglie wavelength* of a particle of mass m with momentum $p = mv$ is:
>
> $$\lambda = \frac{h}{p} \tag{1.9}$$
>
> Let's assume a person with a mass $m = 100$ kg is traveling by car at a speed of $v = 50$ km/h, which corresponds to 13.4 m/s. Then the "wavelength" of this person is: $6{,}626 \times 10^{-34}/(100 \times 13{,}3) = 4{,}77 \times 10^{-37}$ m. For an electron (mass $9{,}1 \times 10^{-31}$ kg moving at this speed, we get: $\lambda_e = 4{,}9 \times 10^{-5}$ m.

Because of their large mass, the momentum of macroscopic objects at low speeds is so large that the corresponding wavelength becomes extremely small.

▶ So we see: light and all particles can behave like a wave and like a particle.

---

[6]Was correctly explained by A. Einstein; for this he received the Nobel Prize.

1.3 What is Light Actually

1.3.4 Spectral Lines in the Spectrum

As already emphasized, astrophysics relies on passive observation. All information about the stars is to be obtained from the light, the electromagnetic radiation, of the stars. Let's go back to quantum physics and look at the structure of atoms. The simplest atom is the *hydrogen atom*. It has a nucleus, consisting of a positively charged *proton* and an *electron*, which is negatively charged and orbits the nucleus. Since the positive charge of the proton equals the negative charge of the electron, a hydrogen atom is electrically neutral. Now one might expect that the electron could orbit the nucleus on any path, similar to, for example, the planets in the solar system. But here again quantum physics comes into play. The electron can only be on certain paths, we speak of *energy levels*. Therefore, we find electrons only at certain distances from the atomic nucleus. For an electron to jump from a lower level to a higher one, the right energy must be supplied.

If we split the light of the stars or other objects with a prism or grating, we often see dark spectral lines. Now we understand their origin. If enough electrons are lifted from lower to higher levels, then this energy is essentially missing in the spectrum, and we see a dark *absorption line*. An electron, which jumps from a higher level to a lower one, essentially gives off energy, we see an *emission line*.

▶ Special feature of quantum physics: Electrons can only orbit the atomic nucleus on certain paths (corresponding to certain distances).

Each transition between levels corresponds to a certain precisely defined energy and thus to a certain wavelength. The hydrogen line H_α arises from the transition from the shell with $n = 2$ to $n = 3$ (Absorption) or from $n = 3$ to $n = 2$ (Emission). The wavelength is 656.3 nm, so it is in the Red (Fig. 1.12).

As an example, we bring up the famous Whirlpool Galaxy M51. The reddish glowing areas in this galaxy are so-called H-II regions where hydrogen gas shines because electrons jump from the shell $n = 3$ to $n = 2$ (Emission) (Fig. 1.13).

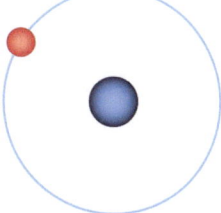

Fig. 1.12 The simplest and most common atom in the universe: The hydrogen atom. It consists of a nucleus (a proton, positively charged) and an electron (negatively charged), which can only orbit the nucleus at certain distances

Fig. 1.13 The Whirlpool Galaxy M51. NASA/ESA

1.3.5 The Doppler Effect

The Doppler effect can be most easily explained with sound waves. Let's assume the following: an emergency vehicle with a switched-on siren approaches a person who is at rest. As it approaches, we hear an increase in pitch, the frequency increases; as soon as the vehicle moves away from the stationary person, there is a decrease in frequency. An increase in frequency means a shortening of the wavelength. This is also outlined in Fig. 1.14.

The same effect occurs with light. As we have seen, light, or more generally electromagnetic radiation, can be described as waves. Therefore, if the wavelength of the light decreases as it approaches an observer, there is a *blue shift,* and a *red shift* when it moves away. The amount of the shift of the light at a wavelength λ depends on the ratio of the object's speed to the speed of light:

Fig. 1.14 Simple explanation of the Doppler effect with sound waves

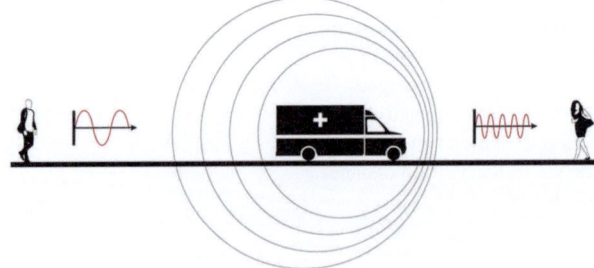

$$\frac{\Delta\lambda}{\lambda} = \frac{v}{c} \tag{1.10}$$

Here, $\Delta\lambda$ is the amount of the shift i.e., the difference between the measured wavelength of the light from a moving source and the laboratory wavelength of a stationary object.

This effect, which is so important for astrophysics but also for other areas, was discovered by the physicist *Christian Doppler* (1803–1853).

1.3.6 Measuring Speeds in the Universe

With the help of the Doppler effect, one can therefore measure speeds, even of galaxies and other objects billions of light years away. The accuracy of the measurements depends only on how accurately one can determine the size $\Delta\lambda$, that is, the wavelength shift. For example, speeds up to cm/s can be determined for the sun.

The astrophysicist *E.P. Hubble* (1889–1953) discovered that (almost all) galaxies are moving away from us. If we compare the velocities of the galaxies with their distances, we strangely find that there is a correlation: the further a galaxy is from us, the faster it moves away from us. This is the famous *Hubble's Law*, which we will discuss in more detail:

$$v = RH \tag{1.11}$$

In this simple relationship, v represents the velocity determined from the Doppler effect, R the distance, and H is the so-called *Hubble constant*. The field of *cosmology* deals with questions about the origin, development, and future of the universe. Here, the redshift is usually given as the so-called $z-$ value. Therefore

$$z = \frac{\Delta\lambda}{\lambda} = \frac{v}{c} \tag{1.12}$$

However, there is a problem: many objects have been found whose z is greater than 1. Let's briefly consider what this means. Suppose we examine a spectral line that is observed at 400 nm under laboratory conditions. Then we get for $z = 1$:

$$\frac{\Delta\lambda}{\lambda} = \frac{\lambda - \lambda_0}{\lambda_0} = z$$

$$\frac{\lambda - 400}{400} = 1$$

and we get: $\lambda = 800$; therefore, we would measure the spectral line of an object with redshift $z = 1$ not at 400 but at 800 nm. We already know: light at 400 nm is in the blue range, light at 800 nm would no longer be visible to the human eye, it is in the IR range. The redshift of spectral lines due to the Doppler effect is shown in Fig. 1.15.

Fig. 1.15 Sketch of the redshift of spectral lines. Below: Spectrum in the lab, above spectral lines of an object moving away from us (redshift). Sketch

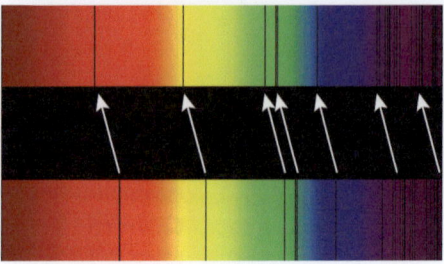

The problem, however, is that according to the Doppler formula $z = v/c$, we would find that the object is moving away from us at the speed of light with a redshift $z = 1$. However, objects have also been found whose redshift is greater than 1, e.g. $z = 4$. This object would have to be moving away from us at four times the speed of light, which of course cannot be the case, as according to Einstein's theory of relativity, the speed of light is the maximum possible speed. For speeds that are very high (from about 10% of the speed of light), one must use the *relativistic Doppler formula*:

$$\frac{\Delta \lambda}{\lambda} = z = \frac{\sqrt{1+v/c}}{\sqrt{1-v/c}} - 1 \qquad (1.13)$$

If we substitute into this formula, for example, $v = 0{,}9c$, i.e., an object moving at 90% of the speed of light, we get $z = 3{,}35$.

Among the most distant objects in the universe are *quasars*. The word means quasi stellar, in normal images they look like stars. However, if their spectra are examined, it turns out that their lines are extremely redshifted. According to Hubble's law, the quasars are therefore very far away from us. In Fig. 1.16 an image of a quasar with a redshift of

Fig. 1.16 Quasars (red arrow) appear like stars when observed with small telescopes. Only the spectrum reveals their large redshift and thus the distance

3.9 is shown. This means that the object is approximately 13 billion light years away from us. Quasars are not brightly shining stars but brightly shining cores of galaxies. The explanation is very exciting. At the center of these cores is a *black hole* with several million solar masses. Matter falls into this, releasing energy. We will encounter black holes and quasars in the following chapters.

Telescopes—Eyes into the Universe

2

In this chapter, we describe how telescopes work, which are the most important tools for astronomers. With them, one can recognize details on the surfaces of relatively nearby celestial bodies such as the sun, moon, and planets. On the other hand, telescopes also collect light, and with them, one can see stars that would not be visible to the naked eye. We will describe the most important properties of telescopes and also give a few tips for those who may be considering acquiring such an instrument for their own observations.

2.1 Basic Types of Telescopes

2.1.1 Lenses

The most important optical elements of a telescope are *lenses* and *mirrors*. What is their task? For simplicity's sake, let's imagine a beam of light. This could consist of many photons lined up. The task of a lens or a mirror is essentially to collect these light rays at a point, this point is referred to as the focal point, *focus* referred. A lens is not flat glass, but the glass is either *convex* or *concave* ground. In Fig. 2.1 you can see a biconvex ground lens. Convex in this context simply means: the lens is thicker in the middle than at the edge. This grind ensures that light rays from the edge of the lens are refracted more than closer to the middle and the light rays are united at a focal point.

In Fig. 2.3 the principle of a biconvex lens is explained in more detail. R_1, R_2 are the radii of curvature of the circularly ground lens surfaces, f is the focal length of the lens, i.e., the distance of the lens's focal point from its center. Such a ground lens collects the light, which is why it is also referred to as a *converging lens*.

In Fig. 2.2 you can see how an object (red arrow) is imaged by a converging lens (green arrow). The image of the object appears enlarged, and it can be easily constructed:

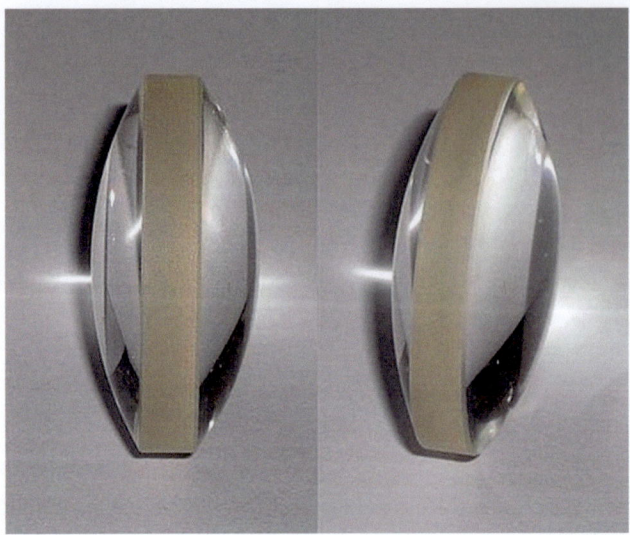

Fig. 2.1 An example of a biconvex ground lens

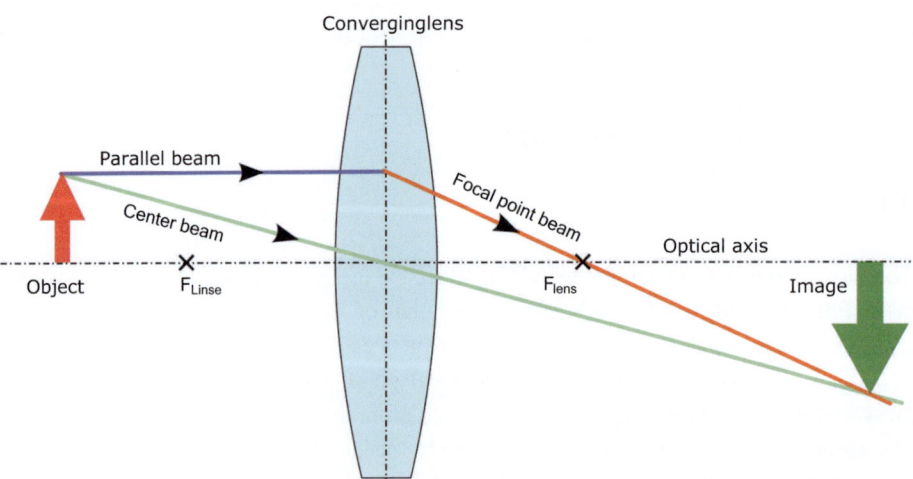

Fig. 2.2 The image of an object (red arrow) can be easily generated from the center ray and a light ray parallel to the optical axis (green arrow)

the light ray that goes through the center of the lens *(optical axis)* is not refracted, a light ray parallel to the optical axis goes through the focal point behind the lens.

In Fig. 2.3 you can see a biconvex ground converging lens.

Lenses can also be ground concave (2.4), then they are thinner in the middle than at the edge. These lenses therefore scatter the light. Both types of lenses are used to correct

2.1 Basic Types of Telescopes

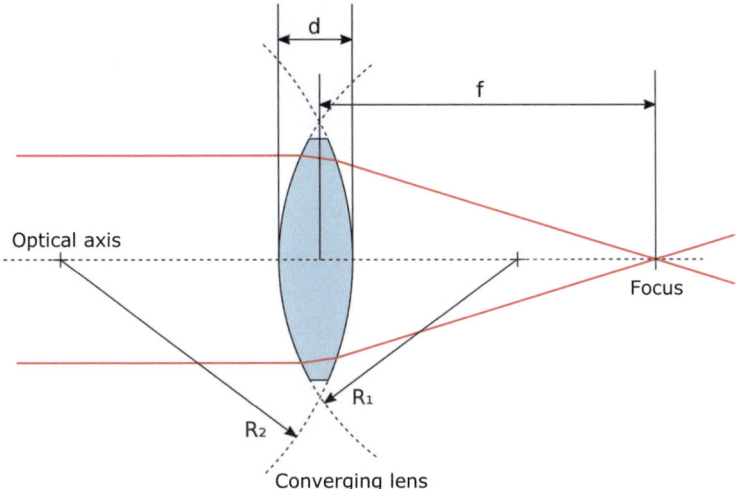

Fig. 2.3 An example of a biconvex ground lens

Fig. 2.4 Principle of a biconcave lens (diverging lens)

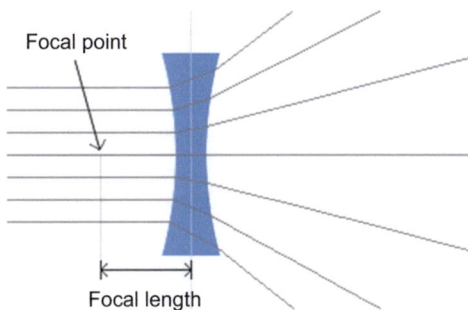

vision defects: Convex lenses are used for farsightedness, the optical center is at their thickest point. Concave lenses are used for nearsightedness, here the optical center is at the thinnest point (Fig. 2.4).

In summary, one can say:

- Converging lens: ground convex, gather light at a focal point.
- Diverging lens: ground concave, scatter light.

In practice, there are different combinations of lenses: biconvex, plano-convex, biconcave, plano-concave, and convex-concave. etc.

The focal length of a lens depends on the refraction of light. The *refractive index* (usually denoted by n) is the most important size here. We all know the following phenomenon. If we look at a straight rod that is held in clear water, it appears broken at the point where it enters the water. This can be easily explained: The refractive index of air is different from that of water.

The refractive index is given by the ratio of the propagation speed of light in a vacuum c_0 to the propagation speed of light in a medium, c_M:

$$n = \frac{c_0}{c_M} \quad (2.1)$$

In a vacuum, light propagates at about 300,000 km/s. So, in one second, it would cover a distance that is more than seven times the circumference of the Earth. In water, the refractive index is about 1.3. The speed of light is therefore only 230,000 km/s.

In matter, light propagates slower than in a vacuum. In near-ground layers of the Earth's atmosphere, light propagates only about 0.02 % slower than in a vacuum; there are certain types of glass where the speed of light is reduced to up to 160,000 km/s.

But now there is a problem that unfortunately makes good telescope optics expensive. The refractive index depends on the wavelength of the light. Short-wavelength light is refracted more than long-wavelength light. What are the consequences? Let's consider the light of a star. This consists of light of different wavelengths. As we have seen, the light of a star can be broken down into its colors (color means wavelength) through a glass prism: we get the colors of the rainbow: violet, blue, green, yellow, orange, and red. Red has a significantly larger wavelength (at about 600 nm) than blue (at about 400 nm). So, if we let a white light beam, which consists of all these colors, pass through a lens, we get a blurry colored image of the star, as the blue component is refracted more than the red component.

This is shown in Fig. 2.5 for blue, green, and red light.

A simple lens, therefore, does not produce good sharp images. This error, also called *chromatic aberration*, can be corrected by using multiple lenses or by specially grinding the lenses. This, of course, makes good optics expensive.

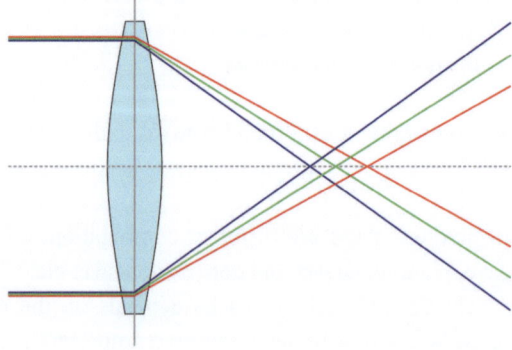

Fig. 2.5 With a simple lens, the lens error known as chromatic aberration results from the different focal points of light of different wavelengths (colors)

2.1.2 Mirrors

Another important optical element in telescopes are *mirrors*. The word mirror, by the way, comes from the Latin: speculum means image. A mirror is a very smooth reflective surface, light is reflected (Fig. 2.6). For a mirror to produce a good image, the roughness of its surface must be less than half the wavelength of the light. Consider, for example, light at a wavelength of 500 nm; a good mirror should therefore have irregularities of no more than 250 nm, which are 250×10^{-9} m. However, when reflecting waves with a wavelength of 1 m, the roughness can be 0.5 m in size.

▶ The larger the wavelength at which observation is made, the greater the roughness of the mirror surface can be.

Mirrors are made, at least in the optical range, by vapor deposition of a thin metal layer (e.g., aluminum, gold, or silver) onto a glass plate. The mirrors we use in everyday life (household, car exterior mirrors, etc.) consist of an aluminum layer behind glass or on plastics. In earlier times, silver layers were used. These tend to tarnish, and the image also gets a yellowish tint. Copper and gold coatings are suitable for infrared, as is the case, for example, with the mirror of the James Webb Telescope.

To collect light through a mirror, it is necessary to grind the reflective surface accordingly. A concave mirror is curved inward. The principle of imaging through a curved mirror is explained in Fig. 2.7. The parallel rays incident on the mirror are focused at the focal point. The reflection of light is independent of its wavelength, so there is no chromatic aberration error here. However, for light rays that are reflected further away from the optical axis, the error of *spherical aberration* occurs. They are reflected at points on the mirror where the curvature is stronger, so their focal point is different. The error of spherical aberration in mirrors can be avoided by not grinding the mirror spherically but in the form of a paraboloid, which in turn increases the cost.

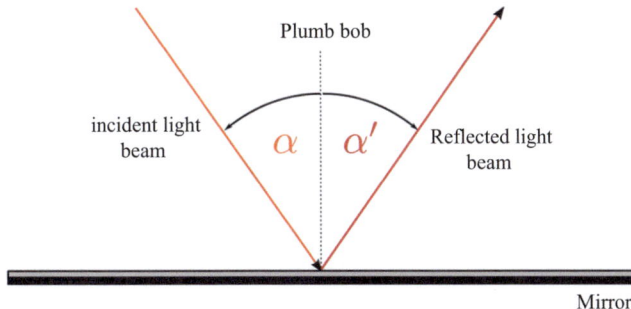

Fig. 2.6 Principle of a mirror. In the reflection of a light beam, the angle of incidence is equal to the angle of reflection

Fig. 2.7 Principle of a curved mirror. The light rays are united at the focal point. (Physik libre, School Physics Book)

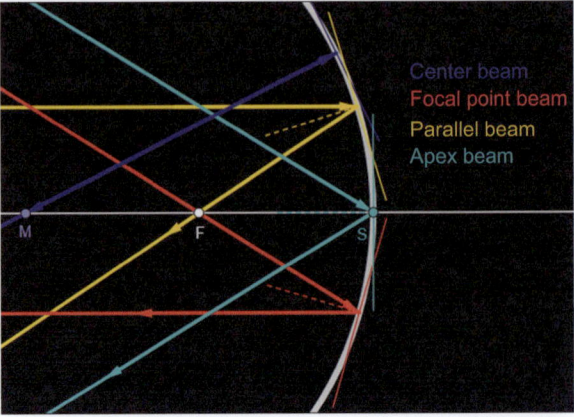

▶ With a parabolic mirror, both rays close to the axis and those far from the axis intersect at the focal point

2.1.3 Lens and Mirror Telescopes

Now it is not a big step anymore to build a telescope. We need the following:

1. a light-collecting surface: this can be a lens or a mirror,
2. an eyepiece, through which one can view an enlarged image.

In a *lens telescope,* also known as a *refractor* , a lens serves as the objective through which the light passes. The object is observed through an eyepiece (Fig. 2.8).

There are advantages and disadvantages to a refractor. The advantage is that one looks in the direction of the object, which makes searching easier. The disadvantage is that

Fig. 2.8 Principle of a lens telescope. The light falls from the front onto the objective (lens). Observation is made through the eyepiece at the rear end. The telescope sketched here has a finder scope on top to facilitate the search for objects

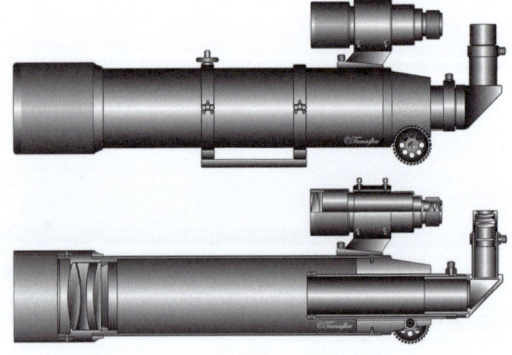

2.1 Basic Types of Telescopes

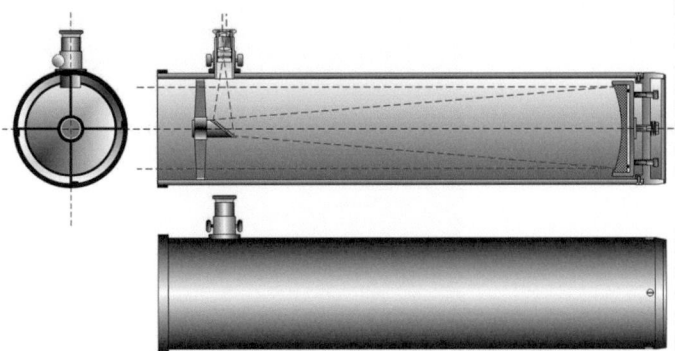

Fig. 2.9 The path of light in a Newton telescope

light must pass through a lens; lenses must therefore be perfectly transparent, which becomes especially expensive for larger lenses. In addition, glass is heavy, and since the lens is mounted at the very front of the telescope tube, it may be that this bends due to the weight of the lens or the lens itself bends due to its own weight. The largest refractors therefore have a maximum lens diameter of about 1 m. As we have seen, lenses have the error of chromatic aberration. Therefore, good telescope objectives must be composed of several lenses to provide color-true images. However, this makes the objective even heavier. Objectives in which this color error is corrected are called *achromatic* or *apochromatic* objectives.

In a *mirror telescope* or *reflector*, a mirror serves as the light-collecting surface. The light falls onto this main mirror, which reflects the light onto a second mirror, and from there it then reaches the eyepiece. There are several types here. In the *Newton telescope,* the view is from the side (Fig. 2.9), so one observes perpendicular to the direction of the object, which makes adjusting the objects very difficult for inexperienced people. This type of telescope was invented by *I. Newton* and can be produced very cost-effectively. A replica of the telescope invented by Newton in 1672 is shown in Fig. 2.10.

In the *Cassegrain telescope,* the light is reflected from the secondary mirror through a hole in the main mirror to the eyepiece, and one observes in the direction of the object. The path of the light is shown in Fig. 2.11. Due to the multiple passage of the light beam as a result of reflection at the main and secondary mirror, these telescopes can be built very compact despite the high focal length of the mirror.

2.1.4 Magnification and Resolution of a Telescope

Generally, the magnification of a telescope is calculated by dividing the focal length of the objective, f_{obj} (mirror or lens), and the eyepiece f_{ok}.

Fig. 2.10 Replica of the telescope invented by I. Newton

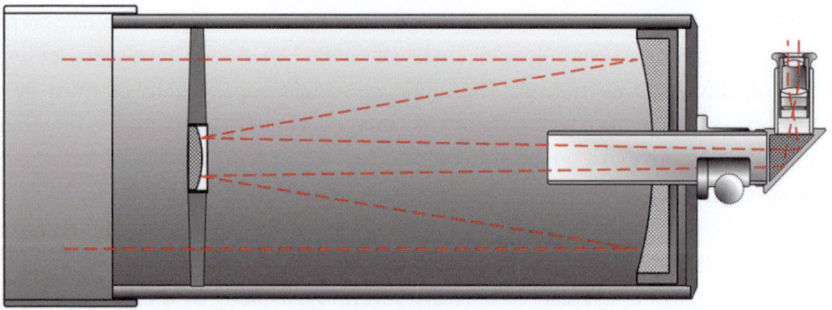

Fig. 2.11 The path of light in a Cassegrain telescope

$$V = \frac{f_{\text{obj}}}{f_{\text{ok}}} \quad (2.2)$$

Let's assume we have a telescope with a focal length of 1 m (the focal length of the telescope usually refers to the focal length of the objective or mirror) and we use an eyepiece with a focal length of 1 cm = 0,01 m. Then this combination would result in a 100-fold magnification. So, we see things 100 times larger than with the naked eye.

Another important characteristic of a telescope is its *resolution*. What do we mean by this term? Telescopes serve to see distant objects magnified, but on the other hand, we want to see not only larger objects but also more details on them. The more details are recognizable on an object, the better is the resolution. This is given by the angle under which, for example, two adjacent points can still be seen separately. Angles are given in degrees, very small angles in arcminutes (') and even smaller in arcseconds ("). An arcminute is 1/60 of a degree, an arcsecond is 1/60 of an arcminute or 1/3600 of a degree. To have an idea of angles in the sky: The diameter of the full moon or the sun in the sky is about 30 arcminutes or 1800 arcseconds. A telescope with a resolution of one arcsecond allows details to be recognized that have an angular extent of one arcsecond. If we were to observe the moon with this telescope, we could recognize details on it that we see from the earth under an angle of one arcsecond. Let's do a rough calculation: The diameter of the moon is 3478 km, which corresponds to about 1800 arcseconds. One arcsecond then corresponds to about 1.9 km. With a telescope which has a resolution of one arcsecond, we therefore recognize details of 2.4 km on the moon's surface.

> The resolution is calculated with the formula:
>
> $$A('') = 1{,}122 \times 206\,265 \frac{\lambda}{d} \qquad (2.3)$$
>
> The resolution is given in arcseconds, λ is the wavelength at which it is observed, d is the diameter of the objective or the mirror. Let's assume we observe in the green wavelength range at $\lambda = 500$ nm, the diameter of the telescope is 10 cm $= 0.1$ m, then the result is: A=1.25 arcseconds.

The resolution of a telescope depends on two factors:

1. Diameter of the lens, mirror: the larger the diameter, the better the resolution, the finer details can be recognized.
2. From the wavelength at which one observes. The larger the wavelength, the worse the resolution.

In Tab. 2.1 we provide some numerical examples.

This table includes two special telescopes: the *James Webb Telescope,* which is located in space, and the *Extreme Large Telescope,* which is still under construction and will be operational at the European Southern Observatory, ESA, in Chile around 2027. The resolution was assumed for all telescopes at a wavelength in the green (500 nm), which is not true in the case of the James Webb Telescope, as it operates in the IR range. The value for the resolution would therefore be at least twice as bad in this case. However, we recognize the following: even with the 40-m telescope, one can only recognize details up to 6 m in size on the lunar surface. The lunar modules left behind by the Apollo astronauts, of which a total of 12 people have set foot on the moon, cannot be seen from Earth even with such a giant telescope.

▶   Telescopes show more details with a larger diameter.

Tab. 2.1 Numerical values for the resolution of telescopes of different sizes and what details can be recognized on the moon with them

Telescope diameter (lens or mirror)	Resolution in "	Details on the moon
0.1 m	1.25"	2.37 km
0.25 m	0.51	0.95
0.5	0.25	0.47
1.0	0.13	0.24
2.0 m	0.06	0.12
5	0.03	0.05
6.5 (James Webb Telescope)	0.02	0.04
10	0.01	0.02
40 (Extreme large Telescope)	0.003	0,006 km = 6 m

2.2 The Earth's Atmosphere

So far, we have discussed the most important types of telescopes and their properties. However, we must consider that when observing from Earth, the Earth's atmosphere is in between and therefore the values given for the resolution in the previous chapter are hardly achieved in practice.

2.2.1 The Permeability of the Earth's Atmosphere

The Earth's atmosphere is only permeable to certain wavelengths; this is referred to as *windows*. Many readers will know that only a small portion of the UV radiation originating from the sun reaches the Earth's surface. The majority of this short-wave radiation is absorbed in the ozone layer of the Earth's atmosphere, and therefore does not reach the surface. *Ozone* is a molecule consisting of three oxygen atoms. The sun's UV radiation initially splits oxygen molecules, which consist of two oxygen atoms (O_2) and the resulting atomic oxygen combines with an oxygen molecule, O_2, and an ozone molecule is formed, O_3. Life on Earth originated in water, as there was no free oxygen in the Earth's atmosphere and UV radiation would have destroyed life on the Earth's surface. Only through *photosynthesis* did free oxygen enter the Earth's atmosphere and an ozone layer could form. Only from this point on could life spread on the continents.

In the higher layers of the Earth's atmosphere, the even shorter-wave X-ray radiation from the sun is also absorbed. Therefore, if one wants to study stars in these wavelength ranges, one must do so with satellites outside the Earth's atmosphere.

2.2 The Earth's Atmosphere

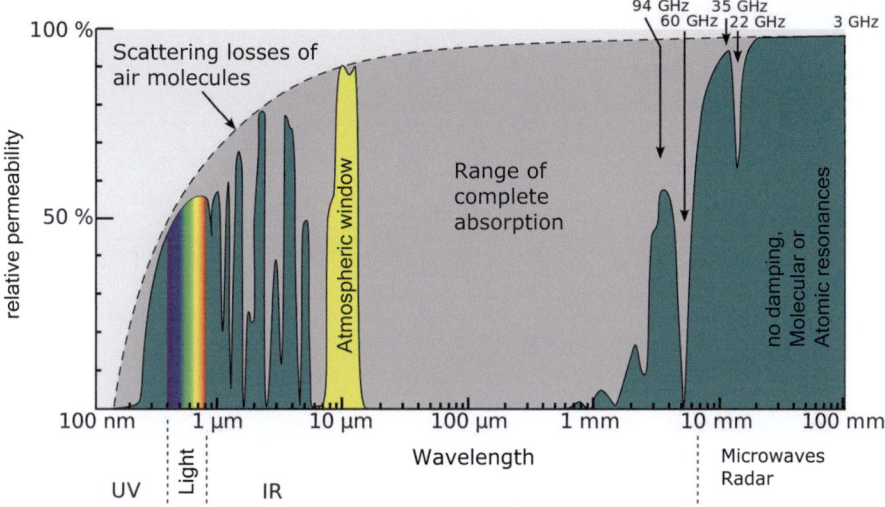

Fig. 2.12 Permeability of the Earth's atmosphere

As shown in Fig. 2.12, there are two permeability ranges of the Earth's atmosphere:

1. *optical window,* in visible light
2. *radio window;* this allows radio communication with satellites and the observation of astronomical objects in the radio range.

The structure of the Earth's atmosphere is shown in Fig. 2.13. Water vapor is primarily responsible for absorption in the near infrared. Weather events occur in the troposphere. Here, hot air masses heated above the Earth's surface rise, cool down, and form clouds, etc.

▶ In the short-wave and in the infrared, the Earth's atmosphere is impermeable.

2.2.2 When do the Sun and Moon Rise?

A simple question, which, as we will see, is not so easy to answer. When we see the sun or the moon exactly on the horizon, strictly speaking, we are not seeing these celestial bodies, but only their elevated image. The effect that leads to this elevation of the image near the horizon is referred to as *astronomical refraction*. As we have already hinted, a ray of light is refracted when it transitions from an optically thinner to an optically denser medium, think of a rod held in water that appears bent at the point of entry into the water surface. This is exactly what happens in the Earth's atmosphere. The air has a slightly higher refractive index than vacuum, i.e., space, from which the light comes.

Fig. 2.13 Structure of the Earth's atmosphere. (Source: Norbert Noreiks, Max Planck Institute for Meteorology)

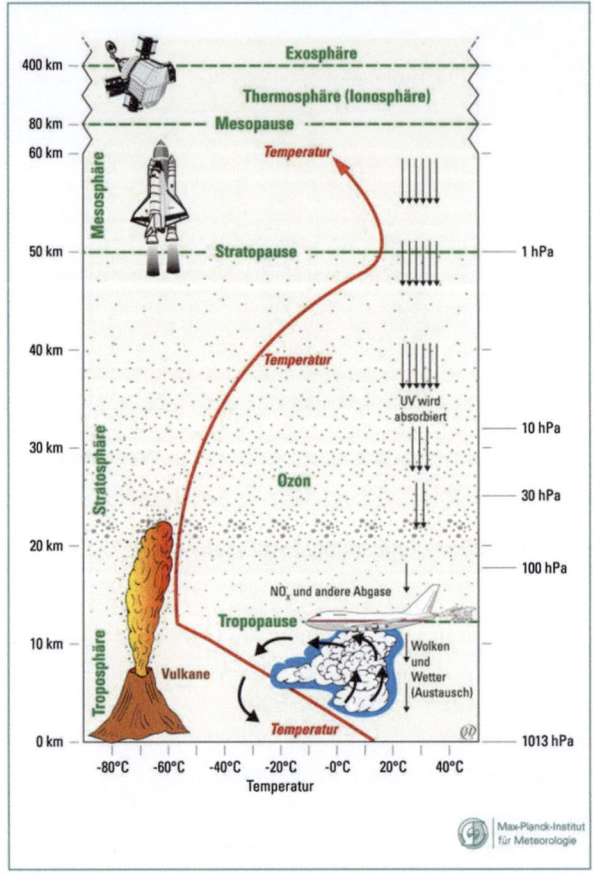

Therefore, light rays are refracted when they enter the Earth's atmosphere, and this refraction is about the apparent diameter of the sun or moon near the horizon. The higher an object is, the less the refraction in the Earth's atmosphere. The refraction of light is a function of wavelength, blue light is refracted more than red. Therefore, the sun and moon appear redder when they are low on the horizon. This effect can also be seen with stars. A very bright star (e.g., Sirius or Vega) appears more reddish near the horizon than at high altitude. The effect of refraction is explained in Fig. 2.14.

In Fig. 2.15, you can see a photograph of the sun at sunrise. As a result of refraction in the Earth's atmosphere, it appears red and distorted, and the image is raised above the horizon, when in reality it is still below it.

Upon closer inspection, we can note: the refraction depends on the refractive index of the air masses, which in turn depends on the temperature, density, etc. Therefore, we see: it is difficult to determine the exact time of the rise of the sun, moon, and other celestial bodies.

▶ Due to refraction, the image of an object in the sky is raised.

2.2 The Earth's Atmosphere

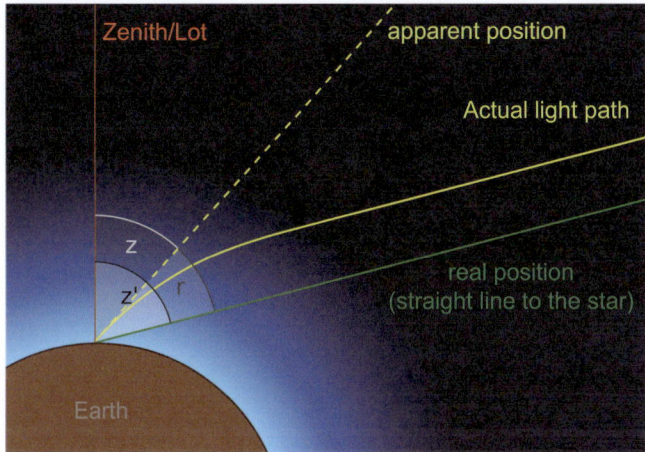

Fig. 2.14 Astronomical refraction: the image of a star appears slightly higher due to the refraction of light rays in the Earth's atmosphere

Fig. 2.15 Astronomical refraction: near the horizon, objects (here the sun) appear red and extended objects distorted

2.2.3 The Turbulent Earth's Atmosphere or Why Stars Twinkle

What distinguishes the starry sky that an observer sees from a spaceship from the starry sky that we see from the Earth's surface? There are two points that immediately stand out: (i) since there are no disturbing light sources in space, the sky appears pitch black, and you see more stars than from the Earth's surface, (ii) the stars appear in space as rigid light sources, they do not twinkle. Especially in winter, where many bright stars appear in the sky for the northern hemisphere, the twinkling of the stars is noticeable. However, for scientific observations, it is rather negative. The twinkling of the stars is caused by turbulence in the Earth's atmosphere. One can imagine the Earth's atmosphere consisting of many layers of air, which have different temperature, pressure, etc. This also changes the light refraction (the refractive index). The consequence is that:

- Stars change their brightness, so they twinkle or glitter. This effect is not as strong with the bright planets Venus, Jupiter, Saturn and usually also Mars, because due to their relative proximity to us, a bundle of light hits the Earth's atmosphere, which is not so strongly influenced by the turbulence in the atmosphere.
- In a telescope, one can see that the stars dance back and forth, the images of planets appear blurred (referred to as *blurring*) and also move *(image motion)*.

▶ The effects of the Earth's atmosphere on the imaging of stars are summarized as seeing.

Due to the "seeing", the astronomical objects in the telescope do not appear sharp but blurred. Good seeing is spoken of when the effects are less than one arcsecond. As already shown, a telescope with about 10 cm diameter provides details up to an angle of one arcsecond. Due to the seeing, therefore, one cannot recognize more details with a telescope of 10 m diameter than with a telescope of 10 cm aperture. Telescopes beyond the 1-m limit quickly cost several million euros. Why then this effort? An important property of a telescope is to collect as much light as possible to see weaker objects. In Fig. 2.16 you can see images of the planet Jupiter at different seeing conditions.

2.2.4 The Human Eye: Our Natural Detector

The images of a telescope must be stored. Modern cameras no longer rely on photographic film material but on *CCDs*. A CCD (charged coupled device) is basically a matrix of light-sensitive picture elements *(pixels)*, which collect the photons. After exposure, the photons stored in each pixel are read out and thus the image is created.

Let's take a brief look at the *human eye*. The light comes through the *pupil* onto a *lens,* which focuses the image onto the *retina*. On the retina, there are two types of light sensors:

2.2 The Earth's Atmosphere

Fig. 2.16 Images of the planet Jupiter at different seeing conditions

- *Cones:* there are three types, which are sensitive to red, green and blue, they enable color vision,
- and *Rods:* they are more light-sensitive, there are significantly more of them, but with them, you only see black/white.

We primarily see with the rods under poor light conditions, hence only black/white. There are about 120 million rods and only 6 million cones in the human eye. A cross-section of the human eye is shown in Fig. 2.17 shown.

The human eye has a pupil as an entrance for light. The size of this pupil depends on:

- Age: the older, the smaller; young people have a pupil with more than 5 mm opening.
- the intensity of the incoming light. In bright light, i.e., during the day, the pupil is smaller (up to 1.5 mm in young people) and larger in a dark environment (night) (up to 8 mm in young people) (Fig. 2.18).

That's why it takes a while when we go from a bright environment out into the dark night before we can recognize weaker stars. This is also referred to as *adaptation* of the eye. It takes about 20 minutes to fully adjust to the darkness. The narrowing of the pupil *(Miosis)* or the enlargement *(Mydriasis)* is controlled by muscles and thus the brightness of the image that falls on the retina, where the light-sensitive organs are.

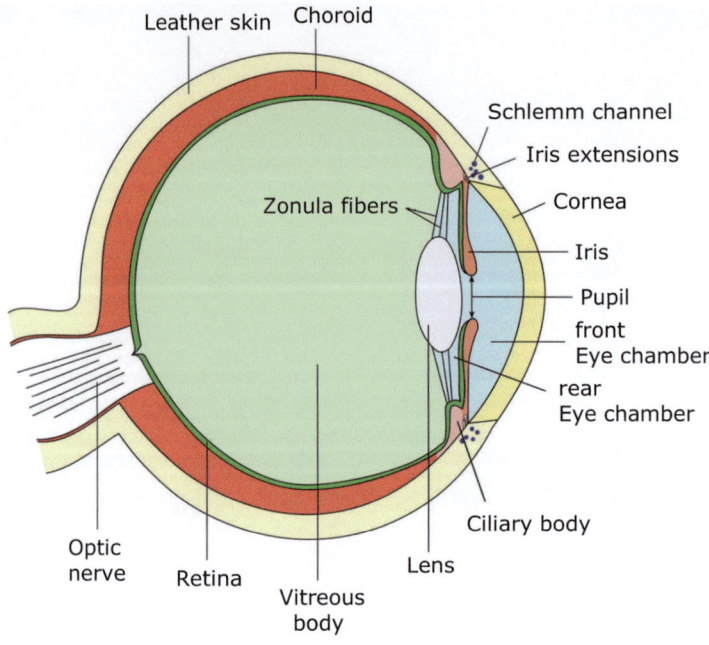

Fig. 2.17 The human eye

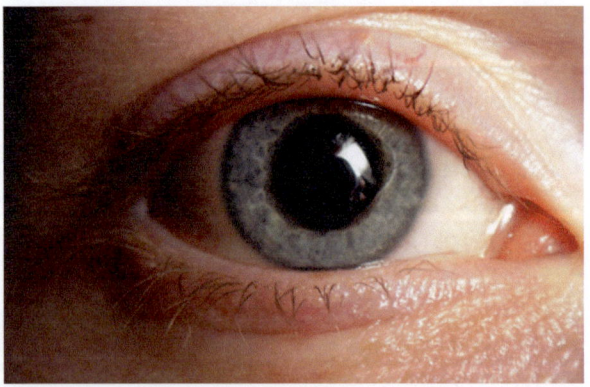

Fig. 2.18 The diameter of the human eye pupil changes and becomes larger under weak light conditions

The larger the opening, the more light is collected. Since the light collecting area corresponds to a circular area, which is given by $A = r^2\pi$ it follows that a telescope with a 20 m opening collects 4 times as much light as a telescope with a 10 cm opening. The diameter of the human eye pupil depends on age, it decreases with increasing age, but let's take the average value of 0.5 cm. The light collecting area is then:

$$A_{\text{Auge}} = 0{,}25^2 \pi = 0{,}0625\pi \tag{2.4}$$

The light collecting area of a 10 cm telescope is:

$$A_{\text{Teleskop}} = 5^2 \pi = 25\pi \tag{2.5}$$

The telescope with a 10 cm opening thus collects 400 times more light than the eye. This allows you to see objects that glow weaker.

2.2.5 CCD

Here, we will briefly discuss modern CCD cameras or CMOS cameras, which have replaced film in astrophysics and photography as a whole.

CCD (charged coupled device) consists of light-sensitive electronic components and were first used around 1969, albeit for data storage. The great advantage, as with film or photographic plate recordings, is that photons can be collected through longer exposure. In this regard, CCD/CMOS cameras are much more efficient than film. More than 50 % of the incident light particles generate a charge, while film only triggers a photochemical reaction in about 1 % (similar to the human eye). Therefore, CCDs or CMOS cameras are much more light-sensitive than film, achieving in 10 minutes with the same shot objects that would require many hours of film exposure. Another advantage is that with a CCD, the recordings are immediately available in digital form and can be further evaluated, whereas films had to be digitized for this purpose in the past.

Fig. 2.19 shows a CCD sensor. The number of picture elements (pixels) in modern sensors is, for example, 4000 by 4000 and often even more.

Fig. 2.19 A CCD

An improvement on the CCD is the so-called CMOS. CMOS stands for complementary metal oxide semiconductor. The sensitivity can be further increased through so-called back-illuminated components. CCD or CMOS chips can be found in every image camera of mobile phones and digital cameras today.

▶ The high light sensitivity of CCD and CMOS now allows images of objects to be taken with amateur telescopes, which in the past were only possible with large telescopes and several hours of exposure with photographic plates.

2.2.6 Photography

Classical photography uses a light-sensitive material (in astrophysics, usually a coated glass plate). Exposure to light triggers a chemical reaction that ultimately produces the image. The sensitivity is determined by the *quantum yield*. How many light quanta must fall on the carrier material for a reaction to occur. Table 2.2 lists the quantum efficiency for the discussed receivers, human eye, CCD/CMOs, and photographic plate. What is quantum efficiency? An efficiency of, for example, 1 % means that 100 light quanta must fall on the detector before a reaction occurs. We can see from the table that CCD/CMOS sensors have more than 50 times the efficiency of the human eye or photographic plate. However, the photographic plate also has a property crucial for astronomy. By exposing for a long time, more and more photons are collected, so weaker objects can be detected.

2.2.7 The View into the Telescope

Many laypeople are very disappointed when they observe galaxies or nebulae through a smaller telescope. Apart from faint gray-white shades, hardly anything is recognizable, especially no colors. We remember how the human eye works. The light-sensitive receptors in the eye are the cones (for color vision) and the rods (for black/white vision). The rods are much more numerous on the retina of the eye and also much more light-sensitive. Only with them can we perceive faintly glowing objects. This is the reason why we see faint objects in the telescope in black and white only. Only longer exposures with the photographic plate or CCD make it possible to recognize the colors of the objects.

Tab. 2.2 Quantum efficiency of different receivers

Receiver	Quantum Efficiency
Human Eye	1 %
Photographic Plate	1 %
CCD/CMOS	> 50 %

2.3 The Brightness of Stars

2.3.1 How Bright Do Stars Actually Shine

When we look at the starry sky, we see very bright stars, stars that shine a little less brightly, and finally stars that can just be perceived with the naked eye. Many laypeople believe that we see millions of stars in the sky, but under favorable conditions away from light pollution and in complete darkness, there are fewer than about 3000 stars that can be perceived in the sky. In antiquity, the brightness of the stars was given in *magnitude classes*, the Latin term is *magnitude*. The brightest stars were called first magnitude stars, written as 1^m, slightly dimmer stars were referred to as second magnitude stars, 2^m and stars that were just visible to the naked eye were called 6th magnitude stars (6^m). This rough scale was then expanded and scientifically precisely defined. Fine differences in brightness are indicated by decimal representation, a star of magnitude class $1^m, 1$ is thus barely perceptibly dimmer to the human eye than a star of magnitude class $1^m, 0$. One can measure magnitude class differences of less than $0^m, 001$. The scale from 1 to 6 was then expanded: stars brighter than 1^m are referred to as 0th magnitude stars. The brightest planet, Venus, is much brighter, it can reach up to $-4^m, 5$ brightness, the full moon about -12^m and the sun about $-26^m, 5$.

Stars change their brightness, which can have various causes:

1. it is a *binary star* and the two components, which orbit around the common center of mass, cover each other.
2. there are large planets passing in front of a bright star, called *transit*. During a transit, there is a very small decrease in the star's brightness.
3. the stars are variable; the group of *Cepheid stars,* for example, are giant stars that inflate and contract, thereby changing their brightness.

These examples are intended to underline the great importance of exact brightness measurements. In the constellation Perseus is the Star, *Algol.* Long it has been known that this star changes its brightness, which can also be seen with the naked eye. In Fig. 2.20 it is sketched how the brightness of the entire system, which consists of two stars, changes. From our perspective, there is an eclipse. The first deep minimum (2) occurs when the smaller cooler star moves past the brighter star from our perspective, the second smaller minimum when the smaller star disappears behind the larger one (4). If you see the two components, the light curve is constant and the brightness is greatest (around 1 or 3).

The star is 93 light years away from us, the minimum lasts about 5 hours and the period of the light change about 2.86 days. If you observe this star in the radio range, you also see strong brightness outbreaks. This is explained by a mass exchange

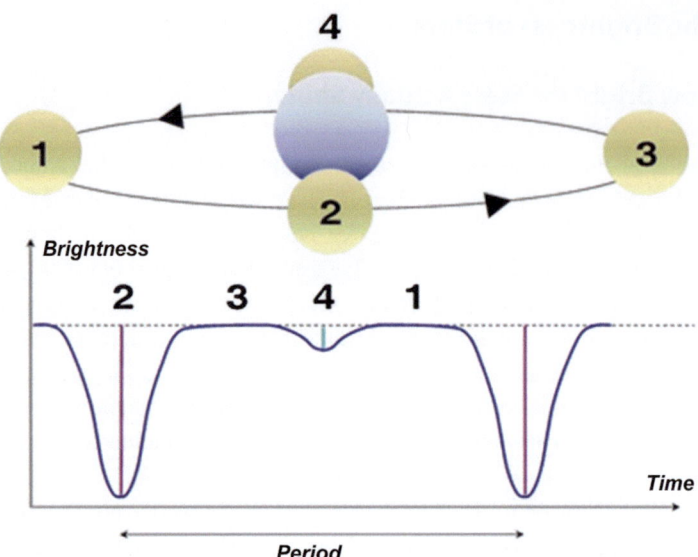

Fig. 2.20 Explanation of the brightness change of a so-called eclipsing variable star, here as an example the star Algol

between the two stars. The two stars are only 0.062 AU apart. An astronomical unit, AU =150,000,000 km, the average distance Earth-Sun. The distance between the two stars is therefore only 10 million km, 30 times the distance Earth-Moon.

The brightness of a star is proportional to intensity, the definition is:

$$m = -2{,}5 \log I \qquad (2.6)$$

Two stars that differ in their brightness by one magnitude differ in intensity by 2.512 times. With two magnitude differences, the difference in intensity is $2{,}512^2$ and so on.

2.3.2 The Absolute Brightness

The brightness thus indicated is also called the *apparent brightness*. It depends on the true brightness with which a star shines and its *distance*. Our Sun is the nearest star. However, it is only so bright in the sky because it is so close to us. The next star is about 300,000 times further away from us than the sun.

To compare the brightness of stars, the term *absolute brightness* has been introduced. This refers to the apparent brightness of a star if it were 10 pc (parsec) away from us,

2.3 The Brightness of Stars

10 pc = 32,6 light years. We recall the method of determining the distances of stars from their annual parallaxes. If the annual parallax is 0.1 arcseconds, this corresponds to a distance of 10 pc. Our sun would hardly be visible to the naked eye at a distance of 10 pc, only under very good conditions; it would be a tiny star of magnitude $4^m, 5$. The absolute brightness of a star is denoted by the letter M, so the absolute brightness of the sun would be $+4^M, 5$.

▶ The larger the telescope aperture, the weaker the objects that can be detected with it.

The following formula can be given for calculating the limiting magnitude, thus the brightness of a star that can just be seen in a telescope:

$$m_{gr,tel} - m_{gr,Auge} = 5 \log \left(\frac{D_{tel}}{D_{Auge}} \right) \qquad (2.7)$$

An example: Let's assume that stars up to a limiting magnitude of $m_{gr,Auge} = 6^m$ are visible to the naked eye. The diameter of the eye pupil is 6 mm, With a 20-cm telescope, stars up to the 13.6th magnitude can then be seen. This value can be increased by about 5 magnitude classes through long exposures with a CCD.

▶ The limiting magnitude refers to the apparent brightness of the weakest star that can just be seen in a telescope.

Some examples of limiting magnitudes are given in Tab. 2.3 (as evident in the above formula, one must specify the diameter of the eye pupil, we have assumed a value of 6 mmm). The table clearly shows that with a good pair of binoculars, one can see significantly more stars (up to the 10th magnitude) than with the naked eye.

Tab. 2.3 With a telescope of aperture D, one can see stars up to limiting magnitude M

Telescope aperture D(m)	Limiting magnitude (weakest stars still recognizable)
0.035 (Binoculars)	9.8
0.1	12.11
0.2	13.61
0.3	14.50
0.5	15.60
1.0	17.11
5.0	20.60
James Webb	21.20
10.0	22.11
40 (ELT)	25.12

2.4 Some Examples of Historical Telescopes

2.4.1 From the Invention of the Telescope to the First Large Telescopes

It is still not entirely clear who truly invented the telescope first. What is known for certain: In the year 1608, a Dutch spectacle maker named *Hans Lippershey* (1570–1619, Fig. 2.21) presented a device for magnifying distant objects. He attached a converging lens in a tube and at a distance from it (about the focal length) a diverging lens. In a sensational demonstration, he offered the council of Zeeland an instrument that "would be suitable for seeing at a distance". This took place on October 2, 1608, and just a year later, such telescopes were being sold in Paris. News of these novel instruments also reached Germany and Italy, and *Galileo Galilei* succeeded in replicating this telescope in 1609. Lippershey naturally applied for a patent for his invention, but he did not get it, as both *Jacob Metius* and *Zacharias Janssen* claimed to be the inventors of the telescope.

One of the first to use the telescope for observations in the sky was Galileo Galilei (1564–1642, Fig. 2.22). His telescope consisted of a converging lens as an objective and a diverging lens, it was not possible to achieve very high magnifications with it, the performance of the Galilean telescope (Fig. 2.23 shows a replica) may correspond to that of modern small binoculars.

Fig. 2.21 Portrait of H. Lippershey, who is considered the inventor of the telescope. (From the book de vero telescopii inventore, 1655)

2.4 Some Examples of Historical Telescopes

Fig. 2.22 Portrait of Galileo Galilei

Fig. 2.23 The telescope of G. Galilei (replica). (c) G. Tomsich/Corbis

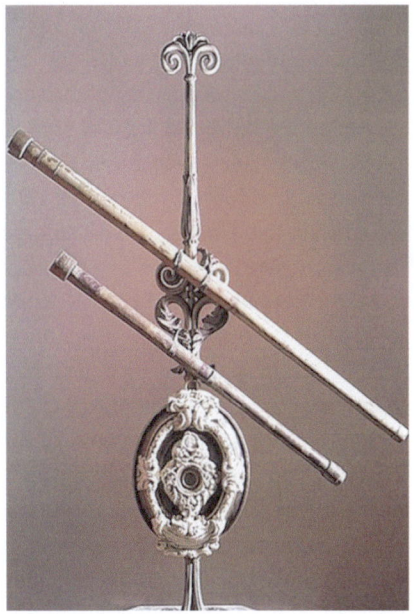

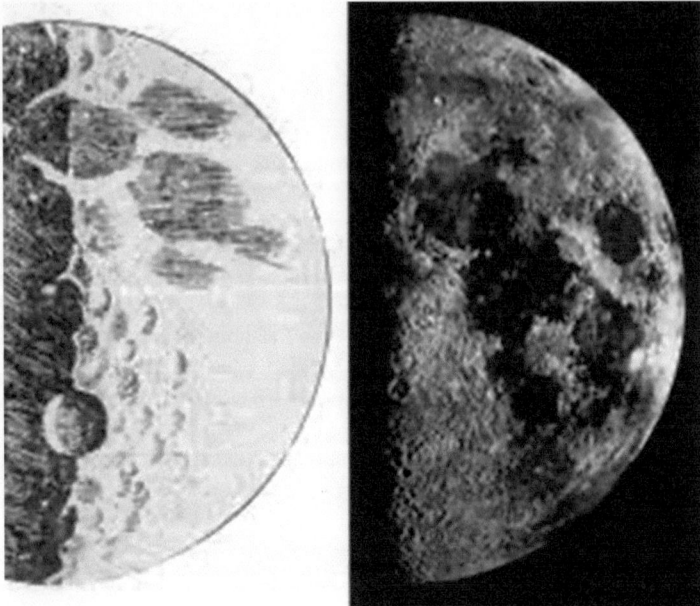

Fig. 2.24 Comparison of the moon map by G. Galilei (left) with a photograph of the moon (right)

But Galilei could see wondrous things in the sky. He found craters and mountains on the moon and created a map of the moon from his observations (Fig. 2.24).

At Jupiter, he found 4 small stars that stood once to the right, once to the left around Jupiter. He concluded that these could be moons of Jupiter. Therefore, these four brightest and largest Jupiter moons are still referred to as *Galilean Moons* today: Io, Europa, Ganymede, and Callisto. Two are larger, two are just smaller than our moon. Galilei saw in these observations the proof that celestial bodies can move around others and was therefore also convinced of the correctness of the heliocentric world system published in 1543 by *N. Copernicus.* Copernicus (1473–1543) said that not the Earth but the Sun was the center of the universe and that the Earth and planets moved around the Sun. This does not initially correspond to our daily experience: Sun, moon, planets, stars rise in the east and set in the west; it would be easier to assume that these objects, firmly attached to the sky, move around the resting Earth on spheres. But Copernicus explains these movements by the rotation of the Earth or the sometimes complicated loop movements of the planets in the sky by the movement of all planets around the Sun. Since the Earth moves faster around the Sun than planets outside the Earth's orbit, they are overtaken by it over the course of a year, which is why there is a retrograde movement of the planets in the sky.

In Fig. 2.25 you can see Galilei's original drawings of the Jupiter moons over several days. The drawings date from the year 1610 as can be read from the handwritten heading. The moons are drawn as small stars around Jupiter. Galilei noticed on consecutive

2.4 Some Examples of Historical Telescopes

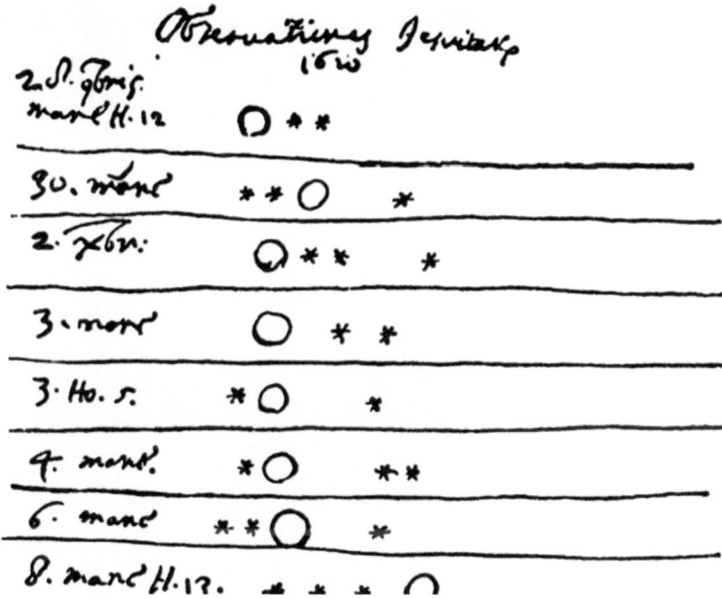

Fig. 2.25 Drawing by Galilei in which he records the movement of the 4 brightest Jupiter moons

evenings how these small stars are always near Jupiter, but their position changes. After a few weeks of recording these movements, he was sure that these were moons of Jupiter that orbit the planet with different orbital periods. The orbits of the four moons are drawn in Fig. 2.26.

Galilei also discovered dark spots on the sun, although it was initially unclear what their nature was. Are they planets passing in front of the solar disk or clouds or are they phenomena of the surface of the sun, as Galilei correctly suspected. However, this brought him into conflict with the church doctrine, according to which the sun had to be a flawless body where sunspots should not occur. Despite his friendship with the Pope, Galilei was eventually sentenced to lifelong house arrest in a villa near Florence.

Sunspots were observed not only by Galilei but also by other astronomers (*C. Scheiner, J. Fabricius* and others). A drawing showing the migration of a sunspot on 10 consecutive days is given in Fig. 2.27. The migration can be easily explained: Sunspots are firmly connected to the surface of the sun and due to the rotation of the sun, they appear to wander around it.

2.4.2 Internship for Enthusiasts: Observing Sunspots

We can easily determine the rotation of the sun from the migration of sunspots. But please be absolutely careful:

Fig. 2.26 The orbit of the four largest Jupiter moons, discovered by Galilei

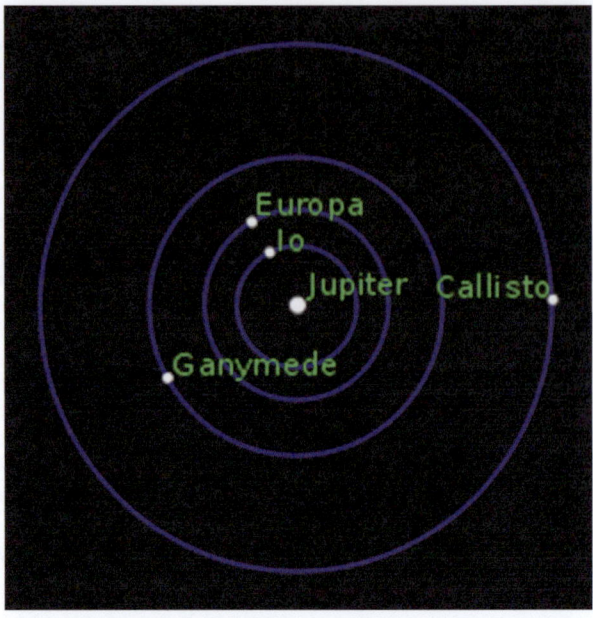

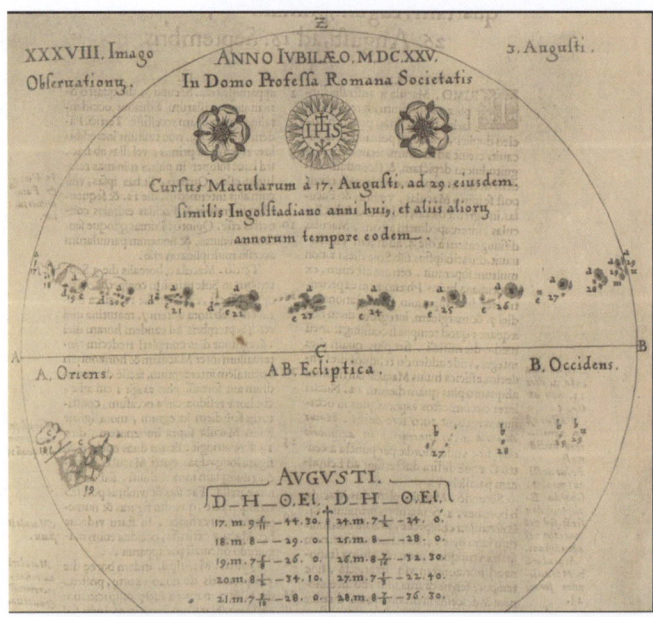

Fig. 2.27 Observations of the positions of sunspots on 10 days. C. Scheiner, 1613

▶ Never look directly into the sun, the retina of the eye would be permanently destroyed, you would be instantly blind!!!

There are sun filters that you can attach in front of the telescope opening (please secure well), then you can safely look into the sun and photograph sunspots with a camera that you attach in place of an eyepiece. Even a mobile phone, properly attached, would suffice. If you compare recordings that were taken over a period of several days, you can clearly see the migration of the sunspots due to the rotation of the sun. However, the activity of the sun and thus the number and size of the sunspots change. During times of minimum activity (about every 11 years) it may be that no sunspots are visible for several weeks or months. The next maximum of solar activity is expected around 2025.

2.4.3 New Planets in the Solar System

Friedrich Wilhelm Herschel (1738–1822) was a German-British astronomer. His goal, when he delved into astronomy around 1770, was to study nebulae and other objects, and it was clear that larger telescopes were needed for this. So he began to build reflecting telescopes himself. Incidentally, Herschel was also a significant musician and became the director of the orchestra in Bath. With his sister Caroline and his brother Alexander, he built new telescopes. In 1781, Herschel discovered a new planet in the solar system, the *Uranus*. This made him famous and he received an annual pension from King George III. Herschel initially named the new planet *Georgium Sidus,* George's Star after the king. This certainly helped to get the annual pension. With the fixed income, Herschel could devote himself entirely to astronomy. Nevertheless, his love for music remained and the composer *Joseph Haydn* visited and admired Herschel's observatory. A representation of the first observation of the planet Uranus by Herschel can be found in Fig. 2.28. Herschel thus used a Newtonian type reflecting telescope.

As already mentioned, Herschel always wanted to recognize weaker objects in the sky and therefore had to build larger telescopes. An example can be seen in Fig. 2.29. This telescope has a mirror diameter of 48 inches, which is 122 cm and had a length of 40 feet, which is 12 m. It was the largest telescope in the world for a long time but was destroyed by a storm in 1839.

One problem with astronomical observations is that the telescope must be pointed with great accuracy at any point in the sky, which of course is a challenge with large telescopes due to their weight. In addition, the observed objects move quickly out of the field of view due to the rotation of the Earth, so the telescope must be tracked according to the daily movement of the objects as a result of the Earth's rotation (Fig. 2.30).

The largest telescope of a similar type was the mirror telescope referred to as the *Leviathan* by *W. Pearsons*. This telescope significantly surpassed the Herschel telescope: the mirror diameter was 183 cm, the focal length 16 m. However, it could only be brought to limited positions in the sky. It was set up in the meridian plane (N-South

Fig. 2.28 Representation of the observation night of March 13, 1781 when Herschel discovered the planet Uranus with his sister Caroline

Fig. 2.29 The 48-inch telescope built by Herschel in 1789 (mirror diameter, 1 inch=2.56 cm). The focal length was 40 feet

2.4 Some Examples of Historical Telescopes

Fig. 2.30 The Leviathan, Birr Castle. The weight of the mirror, which was made of a bronze alloy, is 3800 kg

Fig. 2.31 The DKIST on one of the volcanic islands of Hawaii. NAO/DKIST

direction, so the objects could only be observed when they were near the meridian were), and it could only be moved ± 10 degrees from the North-South axis. The construction time was 3 years and it was put into operation in 1845. Rosse discovered the spiral structure of the M51 galaxy with it. However there was a famine in Ireland at this time, which limited the use of the telescope. The observers had to be free from vertigo, they observed from a gallery at a height of 18 m. The telescope is shown in Fig. 2.31.

This project is one of the last telescopes that were equipped without precise tracking. Tracking is necessary to compensate for the movement of celestial objects caused by the rotation of the earth and to keep the object in the same position in the field of view of the telescope.

The outermost planet in the solar system, *Neptune* was discovered in 1846 based on pre-calculations of its position. It turned out that the orbit of the planet Uranus was disturbed, so there must be another, previously unknown planet involved. U. Le Verrier calculated the position of the new planet from these disturbances and it was then found by J. Gottfried Galle in 1846. However, the planet already appears in Galileo's drawings. In January 1613, there was a close encounter (conjunction) of the planets Jupiter and Neptune in the sky. and Galileo has drawn Neptune as a little star on his drawings.

2.5 Modern Large Telescopes

2.5.1 The Focal Length Determines the Image Size

Modern observatories no longer consist of just a single dome that houses a telescope. A dome serves as protection against weather conditions and also offers some protection against wind; too strong wind would shake the mount of the telescope and the images could be blurred.

Telescopes are used for different purposes.

- to see as many details on objects as possible, a high focal length is needed;
- to see as large a section of the sky as possible, the focal length should be small.

The *focal length* of the mirror or the entrance lens determines the size of the image in the focal plane. There is a simple rule of thumb here: if the focal length is expressed in meters, the diameter of the sun or moon image in the focal plane is the amount in centimeters. So a telescope with a focal length of 1 m produces a 1 cm large image of the sun or the moon.

▶ The focal length in meters gives the diameter of the sun image in cm.

2.5.2 The Largest Solar Telescope

A telescope with a focal length of 1 m thus produces a 1 cm image of the sun or the moon. If one wants to recognize details on the solar disk, for example, it is therefore necessary to use telescopes with the highest possible focal length. The Daniel K. Inouye

2.5 Modern Large Telescopes

Solar Telescope, DKIST (Fig. 2.31) is a solar telescope in Hawaii and the aperture is 4 m. It is currently the largest solar telescope in the world. It was commissioned after about 7 years of construction in December 2019.

In Fig. 2.32 one of the first images taken with the DKIST is shown. It shows an 8000 x 8000 km section of the sun. The cell-like elements are referred to as *granulation*. These are convection currents: hot solar plasma rises from the hot interior of the sun to the surface *(photosphere),* cools down and sinks back down in the dark interstices (inter-granulum). *Convection* plays an important role in the structure of stars. Very hot stars have a convective core, where a constant mixing of elements takes place, cooler stars, including the sun, have a convective shell, which in the case of the sun extends about 200,000 km deep below the solar surface. The image shows details up to 30 km in size. The observatory is located on the slightly more than 3000 m high volcano Haleakala on the Hawaiian island of Maui. The construction costs were about 350 million US$. A total of 22 American research institutions are involved.

The complicated beam path is shown in Fig. 2.33. The light first falls through the aperture onto the 4-m main mirror M1. Then there are further mirrors M2 to M7. The mirror captures a lot of sunlight, which corresponds to a power of 12.8 KW. Through a special device, this light intensity is reduced to about 300 W. The focal length of the main mirror is 8 m.

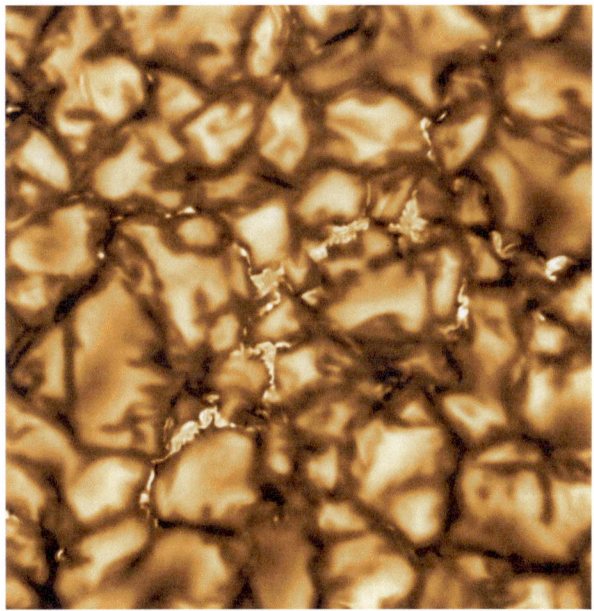

Fig. 2.32 One of the first images of a small section of the solar surface with the DKIST

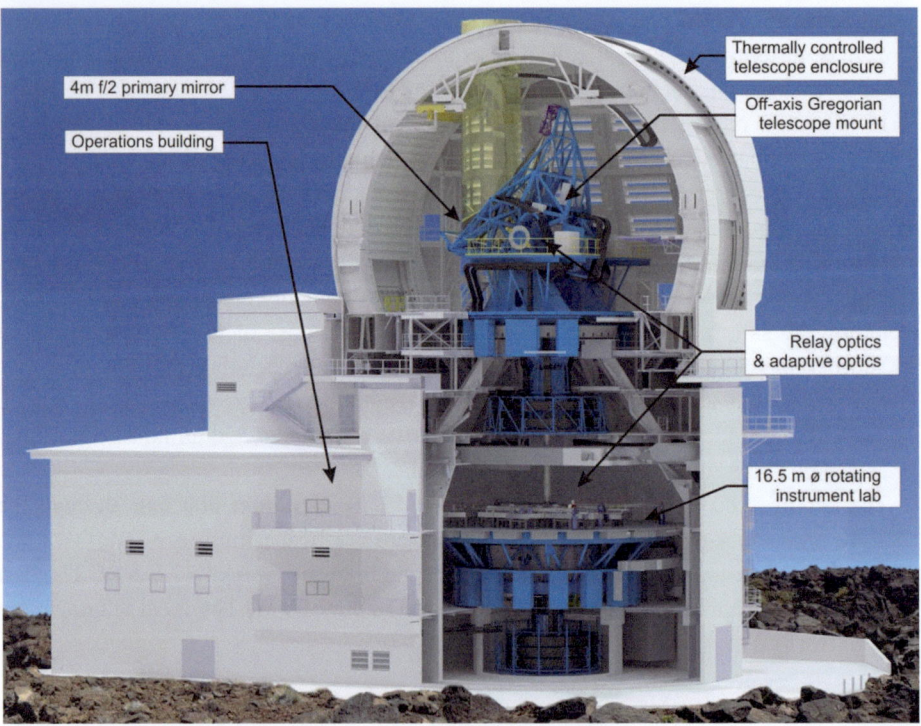

Fig. 2.33 The complicated structure of the DKIST. NSO,/DKIST

2.5.3 The European Southern Observatory, ESO

The European Southern Observatory, *ESO* (European Southern Observatory) is a consortium of 16 European countries that jointly finance this large research facility; the telescopes are located in the Chilean Andes regions. It was founded in 1962, and the headquarters is located in Garching near Munich.

Why did they go to Chile in particular? There are excellent conditions for astronomical observations. Above all, the air is very dry, little water vapor in the air allows observations in the near infrared. The air currents are low, so there is usually very good seeing. The ESO operates four sites:

- The La Silla Observatory on the 2400 m high mountain La Silla,
- The Paranal Observatory on the 2635 m high Cerro Paranal
- Atacama (APEX, Atacama Pathfinder Experiment, for simulating Mars conditions) and the Atacama Large Millimeter Array (ALMA) on a high plateau at 5000 m altitude.
- ELT (Extreme Large telescope) on the 3060 m high Cerro Amazones, completion 2027.

2.5 Modern Large Telescopes

At Paranal, there is the *VLT,* the *Very Large Telescope.* It consists of four telescopes, each with a mirror diameter of 8.2 m. Furthermore, there are four auxiliary telescopes, each with a mirror diameter of 1.8 m. The dimensions of the main mirror are astonishing. It is made of the material *Zerodur;* this is a special type of glass that expands extremely little with temperature changes. Why is this so important? During an observation night, the temperature drops and a mirror made of ordinary glass would deform slightly. Therefore, the focal length changes and the images would become blurry. To keep the weight of the mirror low, despite its 8.2 m diameter, it has a thickness of only 178 mm. Nevertheless, the weight is 23,000 kg. And this weight, and of course much more, must be directed with the highest precision to every point in the sky. The scientists and technicians are accommodated in their own hotel, which is equipped with appropriate protective devices to keep light pollution as low as possible (Fig. 2.34).

The four telescopes can be operated in two different ways:

1. As individual telescopes; each of the four telescopes examines different objects.
2. As an *interferometer,* the term *interfere* means to overlay; the four telescopes observe the same object and the light of all four telescopes is combined. This results in the same performance as a 16-m telescope (Fig. 2.35).

Using the example of the VLT, we show how the disturbing shaking of the images (seeing) caused by the Earth's atmosphere can largely be avoided. For one thing, modern ground-based telescopes are built in climatically particularly favored locations, where the air is very stable and little cloud formation is to be expected, in order to be able to use the telescope as often as possible. In addition, *light pollution* plays a major role. The light of larger cities still disturbs astronomical observations 100 km and more away, the sky is no longer completely dark. If the sky background is too bright, even the largest optics are of no use, you will no longer recognize weak galaxies. Relatively small islands are a good location for telescopes, especially if there are high mountains there.

Fig. 2.34 The ESO Paranal Observatory; in the foreground the hotel complex and laboratory rooms. The VLT telescopes are located on the mountain peak. ESO

Fig. 2.35 The 4 VLT telescopes. ESO

If an observatory is built on high mountains, the layer in which clouds form is usually lower and the telescope location is above the cloud zone. At the VLT, an artificial *laser star,* is used, which is generated in the Earth's atmosphere by reflection from a strong laser beam originating from the telescope (Fig. 2.36). Using sophisticated software, it is then measured how this laser beam is influenced by the Earth's atmosphere and the optics are corrected accordingly. Using servo motors, mirrors are slightly deformed and in this way, sharp images are obtained. These deformations, to compensate for the seeing, occur up to 100 times per second.

An optic that largely restores image quality disturbed by the Earth's atmosphere due to deformations is called *adaptive optics*. The principle is similar to that of an *image stabilizer* built into good cameras, which compensates for the shaking when holding the camera (or binoculars).

In Fig. 2.37, you can see a composite image of the VLT telescopes with the arc of the Milky Way over the course of a night.

2.5.4 Observatories on the Canary Islands

The Canary Islands, especially Tenerife and La Palma, are also excellent observation sites for astrophysics. On *Tenerife*, the *Observatorio del Teide, OT,* was built on a 2400 m high mountain ridge. Originally, it was even considered to build the observatory at an even higher altitude on Teide itself, but this was not realized for various reasons. Firstly, for environmental protection reasons, secondly, the costs would have been significantly

2.5 Modern Large Telescopes

Fig. 2.36 A laser beam leaves one of the four VLT telescopes. ESO

Fig. 2.37 Photomontage showing the course of the Milky Way during an observation night at the VLT site. ESO

higher, and thirdly, there are always degassing events on this volcano, which would not be favorable for either the telescopes themselves or the working people. The telescopes are located at an altitude where the *inversion layer* usually lies below, which is the layer where clouds form. The area is well developed by roads mainly used for tourism. Most of the telescopes stationed there are used for solar observation, the largest solar telescopes are the *VTT, Vacuum Tower Telescope,* the new *Gregor Telescope* both operated by German institutions, and the French *Themis Telescope.* The Gregor Telescope (Fig. 2.38) is an open construction, it is cooled by the wind, the telescope is completely outdoors, the protective roof construction can be completely lowered (Fig. 2.39). With a main mirror diameter of 1.5 m, it is the largest European solar telescope. A consortium consisting of more than 14 European countries is currently trying to realize the construction of a 4.2 m *EST, European Solar Telescope.*

As with other modern large telescopes, Gregor also uses adaptive optics, achieving a resolution of 0.1 arcsecond, which corresponds to approximately 70 km on the solar surface. The functioning of adaptive optics is explained in Fig. 2.40. Light waves that are disturbed by the turbulence of the Earth's atmosphere pass through a beam splitter into a system (Hartmann-Shack sensor) that analyzes these disturbances. A computer calculates the correction signals, which then control the correction elements. The simplest case is the use of a 2-axis tilt mirror. This can compensate for image movements ("tip tilt mirror"). However, better systems control the mirror as a whole, i.e., its surface is deformed with the help of actuators.

Fig. 2.38 The Gregor Solar Telescope. The protective cover can be completely lowered. Leibnitz Institute for Solar Physics, KIS

2.5 Modern Large Telescopes

Fig. 2.39 The Gregor Solar Telescope is completely outdoors during solar observation. Leibniz Institute for Solar Physics, KIS

Fig. 2.40 Adaptive optics, principle. After Ecodeluz

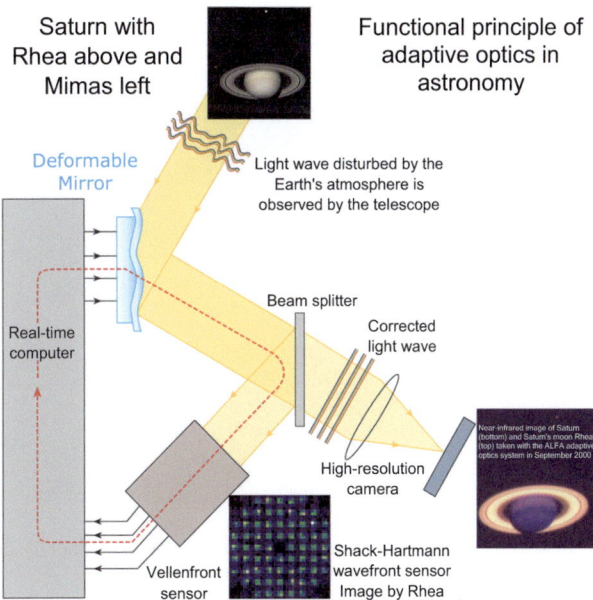

A view of the Observatorio del Teide with its domes is offered by Fig. 2.41. One can see slightly to the right of the center the Vacuum Tower Telescope as the tallest building (70 cm mirror diameter. The telescope optics are located in an evacuated tube to prevent heating by sunlight), the Gregor telescope, and the 1.5-m Carlos Sanchez telescope used for night astronomy (Fig. 2.42).

In contrast to most telescopes on Tenerife, which are used for solar observation, the ORM, Observatorio Roque de los Muchachos located on *La Palma* is mainly responsible for night astronomy. It is also located on a ridge at an altitude of 2400 m. Various countries are involved in its operation. In the picture, you can see on the left the *W. Herschel Telescope* (diameter 4.22 m), 2 solar telescopes (*Dutch Open Telescope, DOT, Swedish Solar Telescope, SST,* diameter 1 m) and others. The largest telescope is the 10.2-m *GTC, Gran Telescopio de Canarias*. In Fig. 2.43, you can see the GTC as well as the two *MAGIC Telescopes* (Major Atmospheric Gamma-Ray Imaging Cherenkov Telescopes). These telescopes do not observe in visible light but measure the Cherenkov radiation in the Earth's atmosphere, which is produced when high-energy gamma rays with energies from 30 GeV to 40 TeV trigger particle showers with particles of the Earth's atmosphere. This results in brief flashes of only a few milliseconds duration. Where does this gamma radiation come from? It is emitted by rotating black holes and neutron stars, and it is also hoped to provide insights into the nature of dark matter. The telescopes each have a diameter of 17 m and are completely in the open.

Fig. 2.41 The Observatorio del Teide from a road frequently traveled by tourists

2.5 Modern Large Telescopes

Fig. 2.42 The Observatorio Roque de los Muchachos, ORM, on the island of La Palma. IAC

Fig. 2.43 The Gran Telescopio de Canarias, GTC (top right). In the foreground, the two MAGIC Telescopes. IAC

2.5.5 Telescopes in Hawaii

Hawaii also offers excellent atmospheric conditions for astronomical observations. The W. M. *Keck Observatory* is part of the Mauna Kea Observatory at the summit of the 4200 m high volcano Mauna Kea. The two identical telescopes, donated by a private investor, each have a mirror diameter of 10 m. In July 2007, they were surpassed in size by the GTC. Keck I was put into operation in 1993, Keck II in 1996. The telescopes can be operated individually or as an interferometer, similar to the VLT. Of course, both telescopes have adaptive optics. The main mirrors, like all large telescopes, consist of several hexagonal elements, in this case there are 36 elements. These are individually suspended and adjustable to an accuracy of 4 nm. In Fig. 2.44 you can see the two domes of the telescopes. Other large telescopes at the Mauna Kea Observatory include the Japanese Subaru Telescope (8 m main mirror) and the *Gemini North Telescope* (main mirror 8 m, operated by USA, Canada, Argentina, Australia, Brazil and Chile).

In Fig. 2.47 you can see a size comparison of some large telescopes that we have discussed or that will be discussed in the following chapters (like the *James Webb Telescope* the *Hubble Telescope* and others). The two Gemini telescopes are identical with a diameter of 8 m. *Gemini North* is located at the Mauna Kea Observatory, *Gemini South* at Cerro Pachon in Chile. Gemini North went into operation in 2000, Gemini South in

Fig. 2.44 The two domes of the Keck telescopes. Keck foundation

2.5 Modern Large Telescopes

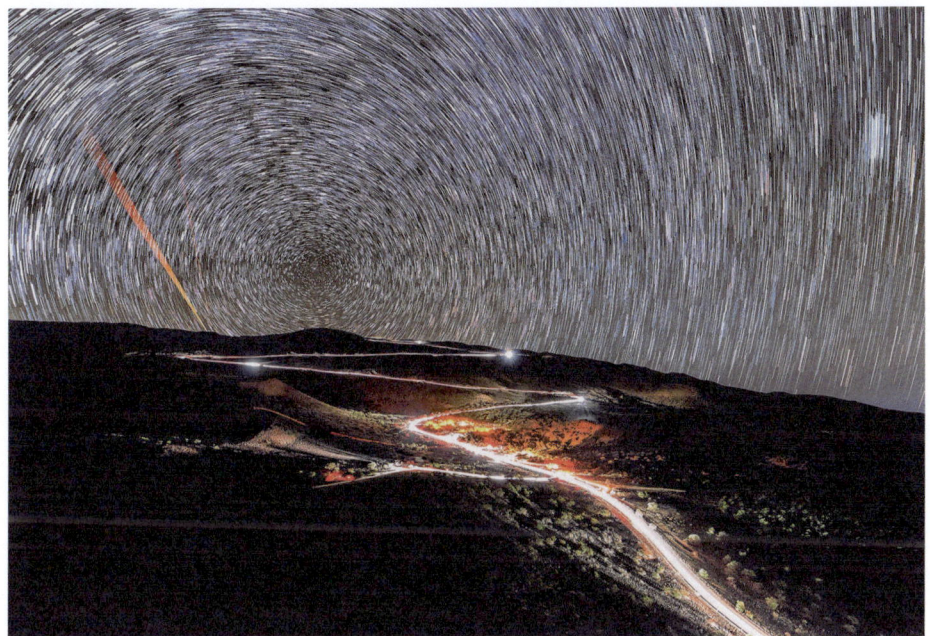

Fig. 2.45 Gemini North in a long exposure shot. https://noirlab.edu/public/images/iotw2147a/

2002. In Fig. 2.45 you can see a night shot of the Gemini North Telescope with the laser beam, which always points in a certain direction, while the stars seem to circle around the celestial north pole due to the rotation of the earth.

In Fig. 2.46 you can see the design of the Subaru Telescope. The weight of the main mirror is 20,000 kg, the thickness only 20 cm (at 8 m diameter!) (Fig. 2.47).

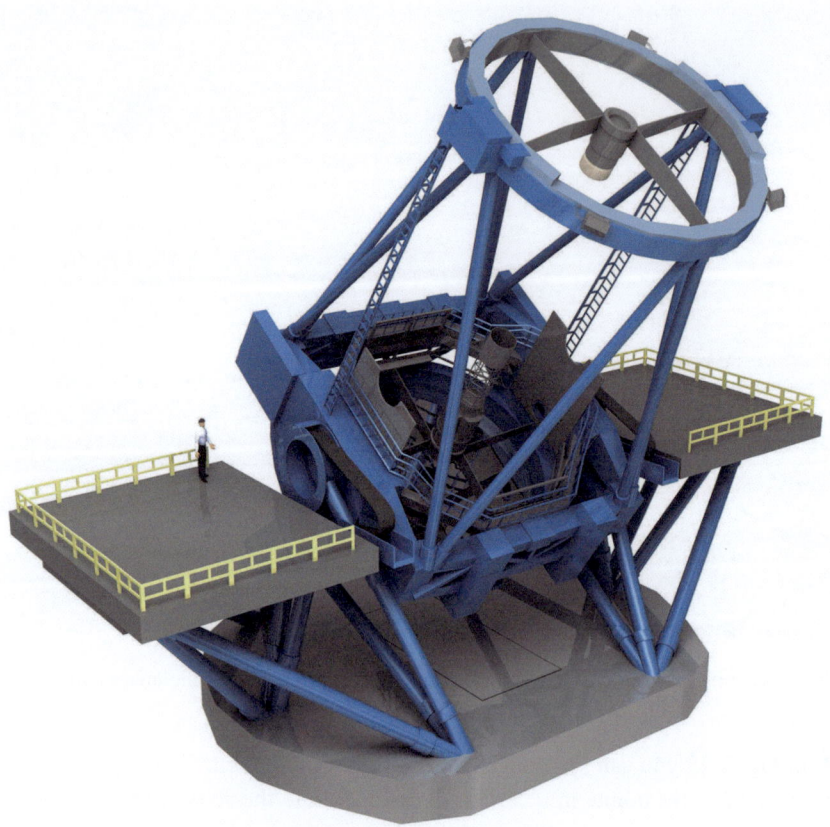

Fig. 2.46 Design of the Subaru Telescope. NAOJ

2.5 Modern Large Telescopes

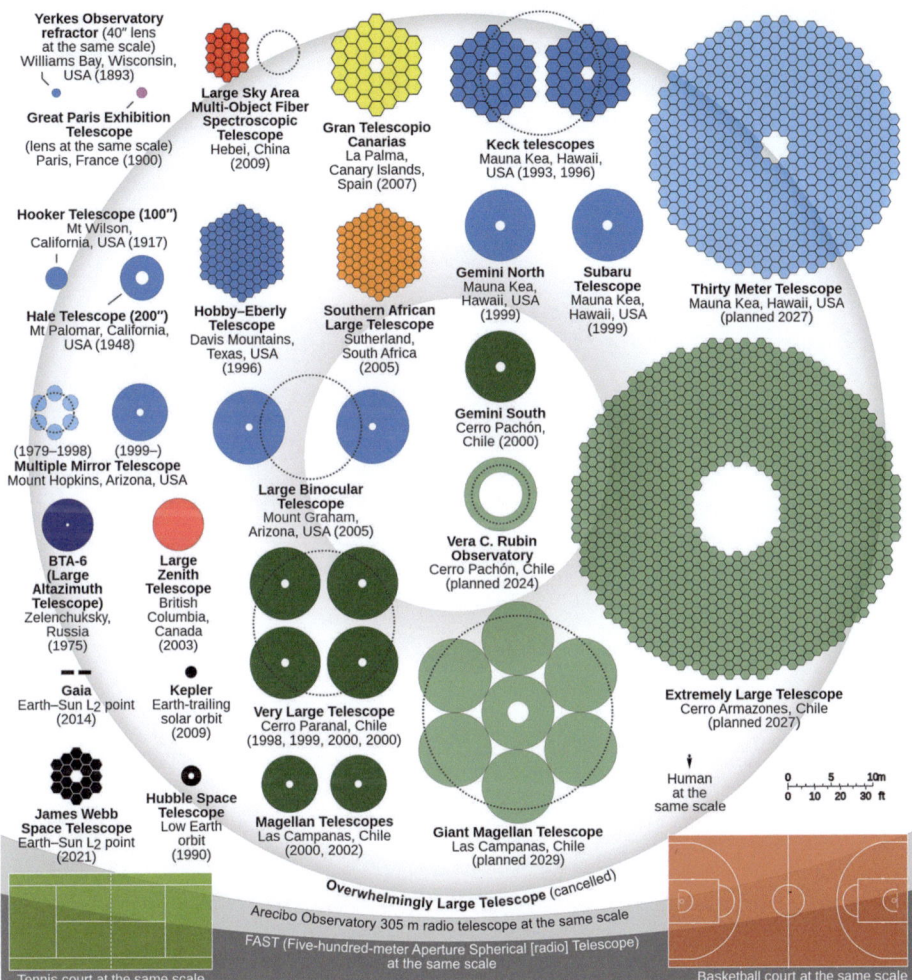

Fig. 2.47 A size comparison of some telescopes; below is the size of a tennis court or a basketball court given. (Source: Extremely large telescope)

Light from the Edge of the Universe—The World of Galaxies

3.1 Size and Expansion of the Universe

3.1.1 Our Milky Way—Home in the Cosmos

The Milky Way, also referred to as the *Galaxy,* is our home in the cosmos. But what does the Milky Way consist of? The ancient Greeks really thought of spilled milk in the sky. The illegitimate son of Zeus, Heracles, was said to have sucked so violently at the breast of Zeus's divine wife, Hera, that milk was splashed across the sky.

Galilei was one of the first to observe the galaxy with a telescope and recognized that the faintly shimmering band dissolves into countless closely spaced stars. The first more precise considerations of what the Milky Way might look like go back to *Herschel.* He made a sketch in 1785 that shows the structure of the galaxy (Fig. 3.1). He based this on star counts. What he did not know: in addition to the many stars, there is also dark dust, and so his representation of the structure of the Milky Way was distorted. But already *Democritus* (died around 370 BC), a Greek philosopher, held the view that the Milky Way is actually composed of countless stars composed.

The position of the Sun in the Milky Way was also known at the beginning of the 20th century. Let's imagine a dark night sky on the northern hemisphere of the Earth. In summer or early autumn, you can see the faintly glowing band of the Milky Way in the evening. However, the Milky Way is hardly visible in winter and spring. This is because we are not in the center of it but about 30,000 light-years away from the center. Over the course of a year, the Earth moves around the Sun. In summer, we look into the inner part of the Milky Way, towards the center. Naturally, the star density is highest there and the Milky Way therefore appears bright to us in summer, the center is incidentally in the constellation of *Sagittarius*. In winter, we are on our orbit around the Sun in the opposite direction to the center of the galaxy and therefore look into regions of the Milky Way

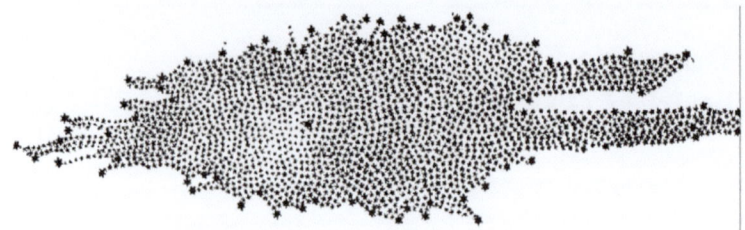

Fig. 3.1 Sketch of the Milky Way made by Herschel in 1785

Fig. 3.2 Our Milky Way seen from the edge: a flat disk with a central thickening. NASA

that are less densely filled with stars. Therefore, from star counts and the distribution of stars in the sky, our place in the Milky Way can be determined. A good introduction to observational astronomy is to observe the Milky Way with binoculars; good binoculars show about a million stars (Fig. 3.2).

Later investigations provided a fairly accurate picture: seen from the edge, the Milky Way is a flattened *disk* with a central thickening, also referred to as a *bulge*. The extent of the disk is about 100,000 light-years. This disk is surrounded by a halo, in which, in addition to individual stars, the globular clusters are located, these are spherical arrangements of several 100,000 stars. And as already mentioned: in addition to the stars, there is also dust and glowing nebulae in the galaxy. Seen from above, our Milky Way would appear like a huge spiral and our Sun is located in one of the *spiral arms*. This is shown in Fig. 3.3.

Fig. 3.3 Our Milky Way seen from above

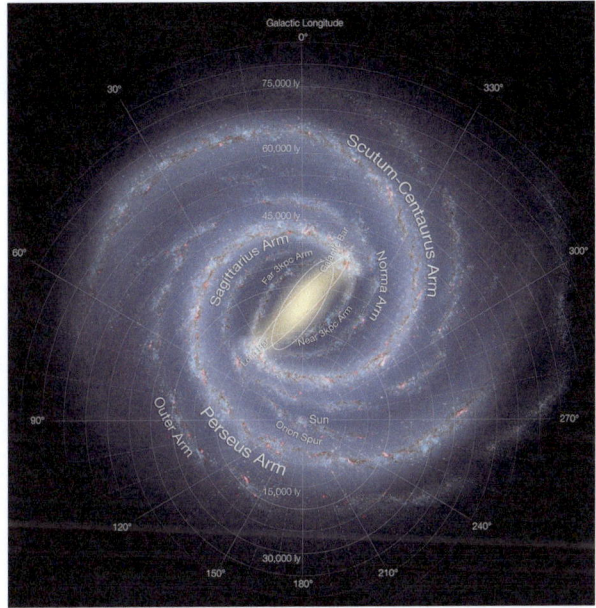

One can get an idea of the size of the Milky Way through the following comparison. Let's assume that the distance between the Earth and the Sun is 1 cm. So 1 cm corresponds to 150,000,000 km. The next star would then be about 4 light years away, which is 40×10^{12} km. In our scale, the distance to the next star would then be 2.6 km. Or in other words, 1 light year corresponds to about 666 m. The diameter of the Milky Way in this model would then be 66,666 km, which is about 1.5 times the circumference of the Earth!

▶ The Milky Way, galaxy, has a span of about 100,000 light years and consists of several hundred billion stars.

3.1.2 Galaxies—Building Blocks of the Universe

At the beginning of the 20th century, astronomy had already advanced considerably, of course modest compared to today's level of knowledge. There was an important tool, photography. With it, celestial events could be captured, positions of celestial bodies could be compared over longer periods of time, and thus, for example, orbits of planets and minor planets could be determined. As already mentioned, despite a similar quantum yield to our eye, photography has a significant advantage over visual observations: the longer you expose, the weaker objects can be seen. This also enables color vision of weak objects. At the beginning of the 20th century, there were already telescopes that could be tracked according to the daily movement of the stars in the sky. Thus, photographic exposures over several hours were feasible. One could observe stars, star clusters and nebulae which it was clear belonged to the Milky Way. But there were also strangely appearing spiral or lens-shaped nebulae.

Among the astronomers of the time, there were two different views on the nature of the *spiral nebulae,* which as shown in the previous chapter, could already be successfully observed visually.

- One group considered them to be objects of our Milky Way system. The approximate dimensions of the Milky Way, the galaxy, were known. As already mentioned, the Milky Way is a system with a diameter of 100,000 light years, several hundred billion stars, of which the Sun is one of them. The big question, however, was: are the observed spiral nebulae objects that belong to the Milky Way? Then the universe would have a span of a few hundred thousand light years.
- Another group, however, referred to these nebulae as island universes, as independent galaxies. If this assumption is correct, the universe must be much larger than our Milky Way.

But how can one determine whether there are other galaxies besides our own, whether the observed spiral nebulae and other small nebulous spots in the sky are independent galaxies? Quite simply: one determines their distance. This task was undertaken by the astronomer *E. P. Hubble* (1889–1953, Fig. 3.4). He was able to work with the then largest telescope in the world, the 2.5-m Mt Wilson telescope (Fig. 3.6).

The object of investigation for Hubble was the Andromeda Nebula, already known to the Arabs. This is located in the constellation Andromeda and can be seen as a faint nebulous spot under good conditions with the naked eye become.

Fig. 3.4 E.P, Hubble

3.1 Size and Expansion of the Universe

Fig. 3.5 How to find the Andromeda Nebula in the sky

Fig. 3.6 The 2.5-m Mt. Wilson Telescope with which the expansion of the universe was discovered around 1920 was

A guide to find this object is given in Fig. 3.5. Binoculars already show it very clearly. Observing the Andromeda Nebula with a telescope is rather difficult, as it has a very large extent and therefore only appears as a bright cloud. It was therefore clear that a large telescope is needed to recognize details in this cloud.

A telescope with a mirror diameter of 2.5 m also shows the weaker outer parts of the Andromeda Nebula and it was even possible to observe individual stars that belong to the Andromeda Nebula. It soon became very clear that the Andromeda Nebula is not a gas nebula in the strict sense, but like our Milky Way, consists of many individual stars. But the final proof was still missing: how far is this object actually from us?

Hubble was able to identify so-called Cepheid stars on the images that belong to the Andromeda Nebula. These are giant stars whose brightness changes periodically within a few days. The really exciting thing about it, however, is that there is a (physically justifiable) correlation between the period of the brightness change of the *Cepheids* and their true (absolute) brightness. Therefore, if you know the easily measurable period, you know how bright the star actually shines. What does this now bring us in the question of distance determination? As we have shown, the apparent brightness of an object in the sky depends on its true brightness and its distance. So if we know—as in the case of the Cepheids—the true brightness, the distance follows from the apparent brightness, which is also easily measurable by comparison with other stars. This is how Hubble was able to determine the distance of the Andromeda Galaxy determine (Fig. 3.6).

Figure 3.7 shows one of Hubble's original plate recordings on which he marked Cepheids in the border areas of the Andromeda Galaxy.

▶ From the period of brightness change of the Cepheid stars, their distance follows

Hubble found a distance of 700,000 light years for the Andromeda Galaxy; the actual value is 2.5 million light years, because Hubble used a wrong relationship between the period and the luminosity of the Cepheids. However, one thing has become clear: The universe is much larger than our Milky Way and there are many galaxies.

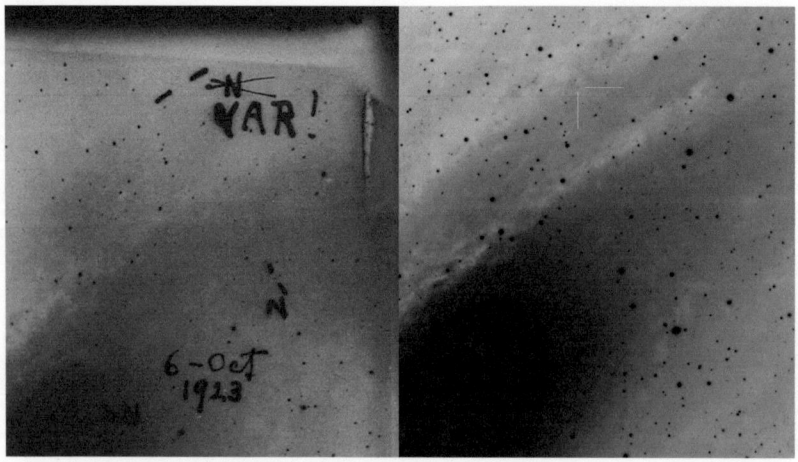

Fig. 3.7 Cepheid stars found by Hubble in 1923 in the outer parts of the Andromeda Galaxy

3.1 Size and Expansion of the Universe

Let's do a small calculation. Let m be the apparent brightness of a star and M the absolute brightness, then it can be shown that

$$m - M = -5 \log d - 5 \tag{3.1}$$

applies. Here, d is the distance in parsecs (1 pc = 3,26 light years). The period-luminosity relationship for Cepheids is:

$$M = -2,81 \log P - 1,43 \tag{3.2}$$

Here, P is the period in days. This is, as mentioned, an empirical relationship, but it can be physically justified. Let's now calculate the absolute brightness of a Cepheid that has a period of 5.37 days (this is the period for the star δ Cephei, the prototype). We then find

$$M = -2,81 \log(5,37) - 1,43 = -3,48 \tag{3.3}$$

This star would therefore shine much brighter than Jupiter in the sky as seen from Earth at a distance of 10 pc = 32,6 light years. But is it possible to see such stars at the distance of the Andromeda galaxy with a 2.5-m telescope? For this, we calculate the apparent brightness and set the distance of the Andromeda galaxy to the value 650 kpc.

$$m = M + 5 \log d - 5 = -3,48 + 5 \log(650000) - 5 = 20,58 \tag{3.4}$$

We see: the apparent brightness of the star δ Cephei would be +20.58 magnitudes at the distance of the Andromeda galaxy. However, such faint stars can indeed be captured with a 2.5-m telescope with a sufficiently long exposure.

▶ Galaxies are thus the building blocks of the universe and consist of up to several 100 billion stars

The Andromeda galaxy, captured with a short focal length telescope and low magnification, is shown in Fig. 3.8. The astronomer Messier created a catalog of just over 100 objects in the sky: galaxies, gas nebulae, star clusters. In this catalog, the Andromeda galaxy is designated M31. Ch. Messier (1730–1817, Fig. 3.9) was interested in the discovery of new comets and since these initially appear as a small nebulous spot but move further from day to day, it seemed sensible to him to create a catalog of objects that do not move further in the sky. Due to the telescopes available at the time, Messier could only include relatively bright objects in his catalog, and therefore these objects are still very popular among amateur astronomers today, as they can be found with small telescopes. Messier worked at the famous Bureau de Longitudes in Paris and observed more than 20 comets.

Fig. 3.8 The Andromeda Galaxy, M31

Fig. 3.9 Charles Messier around 1770

3.1.3 The Expansion of the Universe

Hubble has thus shown that galaxies are building blocks of the universe. But that was not his only great discovery. He also measured the velocities of the galaxies. We remember the Doppler effect. If a light source, in this case a galaxy, moves away from us, then its spectral lines in the spectrum are shifted to red. The greater the redshift, the faster the

3.1 Size and Expansion of the Universe

galaxies are moving away from us. To record a spectrum of a galaxy, one must therefore break down its light through a glass prism or grating and then analyze the lines. To his great astonishment, Hubble found that almost all galaxies are moving away from us. Are we therefore the center of the universe and everything is moving away from us? Furthermore, Hubble had another brilliant idea. He plotted the measured redshifts or velocities of the galaxies moving away from us against their distance, and a simple correlation emerged:

▷ The greater the distance of a galaxy, the greater its speed at which it moves away from us.

This leads to the *Hubble Law:*

$$v = dH \tag{3.5}$$

v is the velocity, d the distance of a galaxy and H is a constant, the *Hubble constant.* The original data from Hubble or Lemaître (1894–1966), who already suspected a correlation between velocities and distances two years earlier, can be seen in Fig. 3.10. The more distant galaxies one can measure, the better the law is depicted. In the figure, galaxies up to a distance of 2.5 Mpc (that's 2.5 million pc, 1 pc = 3,26 light years) were considered (Hubble). The Hubble constant simply results from the increase of the imaginary straight line through the points.

Lemaitre was a priest and scientist and he held the view that the universe was expanding, but his works appeared in 1927 in a rather unknown scientific journal, so today the discovery of the expansion of the universe is generally attributed to Hubble. The fact that the galaxies are moving away from us can be easily understood with an expanding universe. In an ever-expanding universe, all galaxies are moving away from each

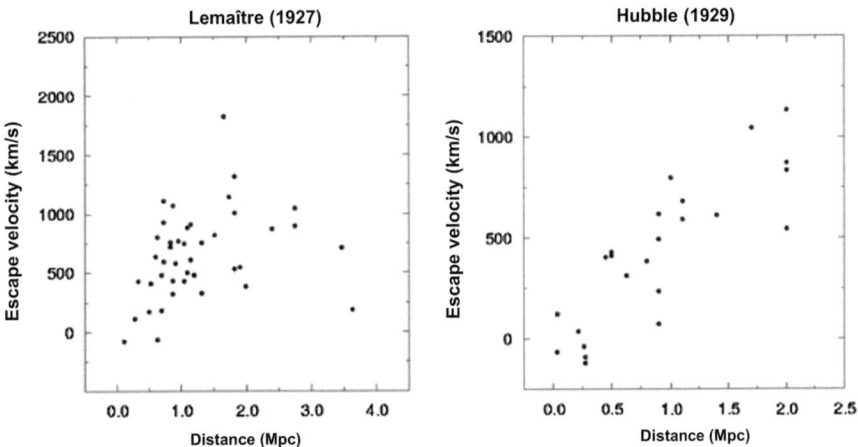

Fig. 3.10 Original data used by Hubble and Lemaitre

other. So we are not the center of the universe or are in an outstanding place in it. Our colleagues—if they exist—would find the same from any galaxy: all other galaxies are moving away.

▶ All galaxies are moving away from us, but we are not the center of the universe, it is expanding.

3.1.4 The Universe Had a Beginning

We have seen: The universe is expanding. This means it is getting larger over time. It is important to note that this effect only operates on very large scales; on smaller scales, gravity prevails. The expansion of the universe does not increase the distance between the Earth and the Sun, or the Sun and the galactic center. Galaxies arrange themselves into *clusters of galaxies.* Our Milky Way is part of the so-called *local Group,* which consists of about 30 objects, of which the Milky Way and the Andromeda Galaxy are the largest objects. The other members of the local group are dwarf galaxies consisting only of a few million stars. Even within such galaxy clusters, the distances do not increase as a result of the expansion of the universe. However, on larger scales, from about 50 million light years (Fig. 3.11).

But if the universe becomes larger on large scales in the future, it must have been smaller in the past or a point must be reached, if we go further back into the past, where the universe was limited to a tiny point. From this point, everything has developed. This is essentially the *Big Bang theory.* Everything originated from a point (mathematically speaking a singularity).

Fig. 3.11 Lemaitre (center) with Einstein (right) and Milikan (left), 1933 at MIT

Can we calculate the *age* of the universe? The answer is quite simple. We know how fast the universe is expanding, so its age follows from the rate of expansion. This is also seen in Hubble's law. The reciprocal of the Hubble constant corresponds to the age of the universe. Modern measurements give a value of about 70 for the Hubble constant kms^{-1}Mpc^{-1}. What does this value mean? A galaxy at a distance of 1 Mpc = 1 Mio. pc = 3,26 Mio. light years would be moving away from us at a speed of 70 km/s. A galaxy that is 300 Mpc away from us, or about a billion light years, is then moving away from us at 21000 km/s. However, this simple calculation only applies to speeds that are less than the speed of light ($c = 300\,000$ km/s).

For this, we need the conversion from Mpc to km: 1 pc = 3.26 light years =3,26 × 10^{13} km, 1 Mpc is a million pc so 1 Mpc = 3,26 × 10^{19} km.

We calculate the age of the universe from the Hubble Law:

$$v = dH \tag{3.6}$$

- We can take the unit of speed as km/s
- Mpc can be used as the unit for distance.
- This results in the unit of the Hubble constant: kms^{-1}Mpc^{-1}.
- Since km and Mpc represent a unit of length, it follows that the unit of the Hubble constant is 1/s or 1/time. Thus, the reciprocal of the Hubble constant is a time, the so-called *Hubble time*.

The Hubble time thus describes the age of the universe, assuming the expansion occurred uniformly.

3.2 The Universe is Getting Colder

3.2.1 Experiment With the Bicycle Pump

Let's imagine we are inflating an empty tire tube with air using a bicycle pump. In order to press the air into the tube, the volume in which the air is located is reduced during the pumping process, and this air enters the bicycle tube. When the piston of the air pump is moved upwards, air is sucked in from the surroundings, etc. One notices that the air pump heats up. Part of this heating is certainly due to the friction of the pump's piston, but the compression of the air also leads to heating. In such an *adiabatic compression*, the gas thus heats up.

A refrigerator works in a similar way. Gas is compressed by a compressor and expands in cooling tubes in the interior of the refrigerator, thereby reducing the temperature, thus cooling the cooling compartment. Therefore, the temperature decreases during an adiabatic expansion.

This is exactly what we can also expect for the universe. As it expands, it gets colder, the temperature decreases. Perhaps one might ask the following question at this point: if the universe is getting colder, where is the energy/heat being released, what is outside? The answer to this is: there is no outside of the universe. The cooling simply occurs through the expansion of space, but the energy/matter contained therein is constant. Heat is nothing more than the movement of particles. At a temperature of $-273{,}15 = 0$ K, there is no more particle movement. This is therefore the lowest possible temperature. However, quantum physics must also be taken into account here. The *uncertainty principle:* states that the product of position and momentum uncertainty has a finite value. So if the momentum (and thus the speed) of the particles is zero, then the position is infinite, but if the position is precisely determined (because the particle is no longer moving), then the speed and thus the momentum is infinite. Therefore, in terms of quantum physics, there is no cooling to 0 K.

The uncertainty principle in quantum mechanics is:

$$\Delta x \Delta p \geq h \qquad (3.7)$$

Δx is the position uncertainty, Δp the momentum uncertainty and h is the Planck's quantum of action $h = 6{,}626 \times 10^{-34}$ Js. Because of the smallness of this size, one notices nothing of the quantum effects in everyday life.

▶ The universe cools down as a result of expansion. Cooling can only go down to a maximum of 0 K.

3.2.2 Looking Into the Past

The astronomer *Olbers, H. W.,*(1758–1840) formulated an interesting paradox in 1823: If the universe is infinite and has existed for all eternity, then the night sky should not appear dark at all. Because no matter where one looks, eventually the line of sight will hit the surface of a star, if the universe is filled with infinitely many stars. The trivial fact, also called *Olbers' Paradox,* that it is dark at night shows us that at least one of the following points must be fulfilled:

1. The universe is not infinitely large,
2. the universe has not existed for all eternity.

One could now object that in addition to the many stars, there is also dust in the universe that absorbs the light of stars located behind it. But if there were an infinite number of stars, they would make the dust glow and the night sky would still be as bright as day.

3.2 The Universe is Getting Colder

▶ The darkness of the night tells us something about the finiteness of the universe!

So often it is very simple observations that allow very profound conclusions!

Here is the text from Olbers' work: *If there are really suns present throughout the infinite space, whether they are distributed at approximately equal distances from each other, or in Milky Way systems, their quantity will be infinite, and the whole sky should be as bright as the sun. For every line that I can think of drawn from our eye will necessarily hit a fixed star, and so every point in the sky should send us fixed star light, that is, sunlight*

To illustrate Olbers' paradox, consider Fig. 3.12. You see a dense forest. No matter which direction you look, you always see a tree at some distance.

Due to the finite speed of light propagation ($c = 300\,000$ km/s), a look into the depths of the universe is always a look into the past. So if we observe the sun, for example, we see light at the moment of observation that was actually sent from the sun to the earth 8.3 minutes ago. The distance from Earth to the Sun is 150,000,000 km; if you divide this value by 300,000 (speed of light), this gives the time that the light needs from the sun to the earth. The calculation gives $150{,}000{,}000/300{,}000 = 500$ s $= 8.3$ min. The sun is about 400 times further away from us than the moon, so the light from the moon takes just over a second (500/400) to reach the earth. Mars came as close as 90,000,000 km to Earth in December 2022. Let's calculate how long a radio signal from Earth to Mars is on the way. The solution gives 300 s, that's 5 min. Suppose astronauts on *Mars* have a problem and call for help to Earth. The signal is 5 min on the way to Earth and if the call for help is answered immediately, it takes another 5 min. So the astronauts on Mars would have to wait a total of 10 min for an answer, not much help in urgent emergencies. Future Mars missions must therefore run completely autonomously, immediate help and advice from Earth is not to be expected. In addition, the mentioned numerical

Fig. 3.12 Olbers' paradox and a view into a dense forest. No matter which direction you look, you always see a tree

values only apply if the planet Mars is close to Earth. Radio signals in the solar system are therefore on the way from minutes to hours. To the nearest star (*Proxima Centauri*, if you disregard the sun as a star) it takes a radio signal already 4.3 years. Communication therefore requires a lot of patience, we would have to wait more than 8 years for an answer. To the center of the Milky Way it already takes 30,000 years and to the Andromeda galaxy the light is on the way for more than 2.5 million years. So if we observe the Andromeda galaxy this evening, then we see its light that was sent to us at a time when there were just the first ancestors of humans on Earth.

As we will see, with modern telescopes you can recognize galaxies whose light has been on the way to us for more than 10 billion years. The light therefore comes from a time when there was no Earth or Sun yet, because our solar system is only 4.6 billion years old.

▶ A look into the depths of the universe is a look into its past.

3.2.3 Glow From the Time of the Big Bang

The universe has a finite age, which results from the rate of expansion (13.7 billion years). In the early phase shortly after the Big Bang, the universe was extremely hot and cooled down due to the expansion. When the age of the universe was about 400,000 years, its temperature was only around 3000 K. At even higher temperatures, there were only free protons and electrons (and also neutrons). We have discussed the hydrogen atom. It consists of a proton and an electron is bound to it. However, if the temperature of the hydrogen gas becomes too high, exceeding the mentioned 3000 K, then the electron detaches from the proton and there are no neutral hydrogen atoms anymore but only a mixture of free particles. But the photons, the light particles, scatter on these free particles. As a result, the gas becomes opaque. The situation is similar to a fog wall, through which one cannot see, although of course the fog itself consists of small droplets, between which light can indeed pass, but not from a certain distance.

The time when the electrons combined with the protons to form neutral hydrogen atoms is also called the *recombination epoch*.

▶ So we can only look back to the point in the universe's past when it was 400,000 years old.

This is illustrated in Fig. 3.13. The negatively charged electrons form neutral hydrogen atoms with the positively charged protons when the universe was about 400,000 years old, and the photons indicated by arrows could pass through the gas unhindered from that point on. Previously, the photons were scattered by the free electrons.

However, the universe is expanding. As we have seen, an adiabatic expansion of the universe means a cooling down. Therefore, as early as the 1940s, radiation was sus-

3.2 The Universe is Getting Colder

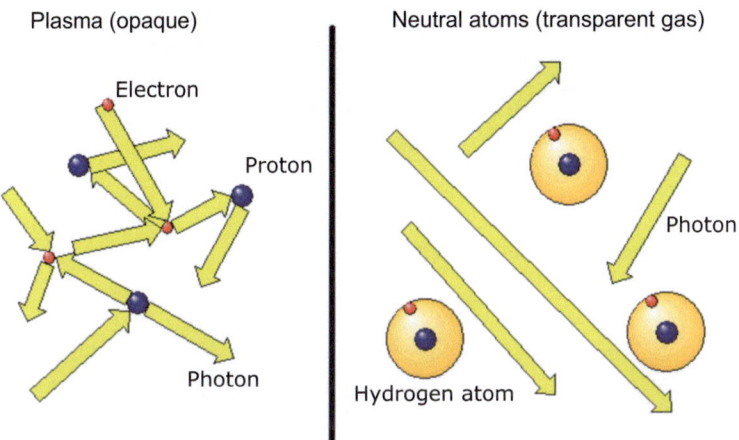

Fig. 3.13 The universe became transparent when the electrons combined with the protons to form neutral hydrogen atoms

pected that could originate from this time, but due to the expansion of the universe, it does not correspond to that of a body at 3000 K but that of a body just above absolute zero. In 1965, a new radio antenna was intended to study the spread of plasma clouds that are ejected from the sun. The sun is active and from time to time there are coronal mass ejections (CME) at which large amounts of plasma are hurled away from the sun. If a radio beam passes through such a plasma cloud, an effect similar to that of the turbulent Earth's atmosphere occurs, the signal scintillates. Therefore, if one observes radio signals from objects in the sky that come from different directions, one can say something about the distribution of these plasma clouds ejected by the sun Plasma cloud (Abb. 3.14).

When the two scientists *A. Penzias* and *R. W. Wilson* began operating their new antenna in 1964, they could only measure noise. Naturally, possible errors were initially considered: problems with the detector's electronics, interference from the environment, contamination of the detector, etc. But systematically, all these sources of error could be ruled out. To great astonishment, this noise was measured with the same strength regardless of where in the sky the radio antenna was pointed. This reminded people of the *background radiation* predicted as early as 1933 by *A. Regener*, a glow from the time when the universe became transparent. Penzias and Wilson had discovered the 2.7 K background radiation and later received the Nobel Prize for it. Later, with the help of a satellite, the intensity of this radiation was measured as a function of wavelength (COBE satellite, Cosmic Background Explorer), and a radiation curve was shown that exactly corresponds to that of a black body with a temperature of 2.7 K. In Fig. 3.15, a comparison between the measured values (COBE satellite) and the Planck curve of a black body at a temperature of 2.7 K is shown. Both curves fit perfectly together!

Fig. 3.14 A coronal mass ejection (CME)

A brief calculation on this. The energy distribution of the background radiation corresponds to that of a black body at a temperature of 2.7 K. At which wavelength or frequency can we therefore expect the maximum radiation to occur? The calculation is simple if you have studied the first chapter carefully. Wien's law states, that

$$T\lambda_{max} = \text{const} = 2897{,}8\,\mu m \tag{3.8}$$

If we first insert the value of the temperature for the sun into the formula ($T_\odot = 6000$ K), we get $\lambda_{max,\odot} = 0{,}482\,\mu m = 480\,nm$. For T=2.7 K, $\Lambda_{max} = 1073\,\mu m$ follows, which is about 1 mm wavelength, thus in the microwave range.

For the frequency, we consider the relationship between wavelength λ and frequency ν:

$$c = \lambda \nu \tag{3.9}$$

therefore, $\nu_{max} = c/\lambda_{max} = \frac{3 \times 10^8 \text{m/s}}{1073 \times 10^{-6} m} = 273\,\text{GHz}$

The cosmic microwave background radiation (English: CMB, Cosmic Microwave Background) is considered important evidence for the Big Bang theory.

3.2 The Universe is Getting Colder

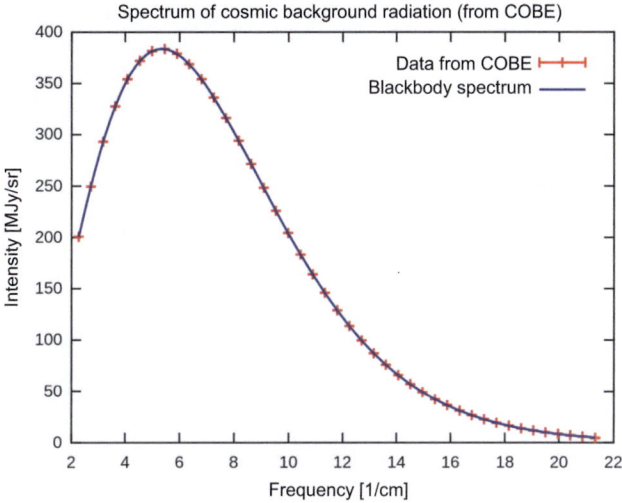

Fig. 3.15 Comparison of the background radiation measurements with the COBE satellite (Cosmic Background Explorer) and the radiation curve of a black body

One could also argue differently. Due to the expansion of the universe, the background radiation has shifted by the factor of redshift z = 1000.

▶ The expansion and background radiation of the universe can be most easily explained with the Big Bang model.

3.2.4 Tracing the First Light

This cosmic background radiation is essentially the first light that we can directly observe. Accurate measurement was only possible with satellites, as we have seen, the radiation lies in the microwave and radio range due to the temperature of 2.7 K.

The first measurements in this wave range were carried out with the *COBE satellite* (Fig. 3.16). This satellite was active from 1989 to 1993 and is still in Earth orbit at a height of 900 km. The different colors indicate different temperatures of the measured areas of the sky, the entire sky is projected onto the map, the Milky Way and other sources of interference have been calculated out. Fig. 3.17 shows the measurement results of the COBE satellite. The top image shows the raw data, which show a clear dipole asymmetry. We are moving together with the Milky Way through the universe, hence light in the direction of movement is blue-shifted, the temperature is therefore slightly higher and against the direction of movement red-shifted, the temperature is therefore slightly lower. If you calculate this effect away, you get the middle image. You can see an irregular bright red stripe across the image. This is the Milky Way in this

Fig. 3.16 The COBE Satellite

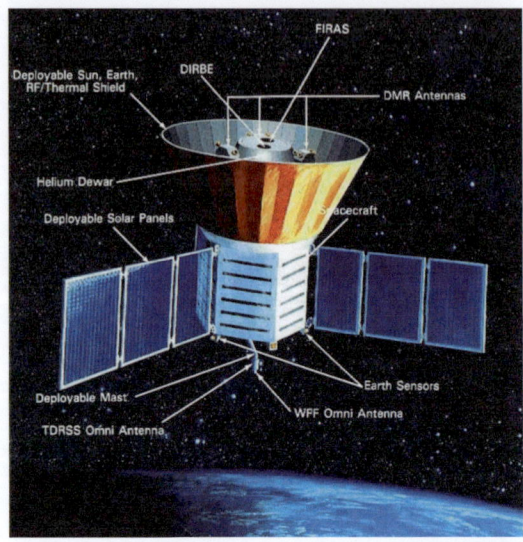

Fig. 3.17 A map of the sky showing very small temperature differences in the microwave background radiation. The data was collected over a period of 2 years. COBE/NASA

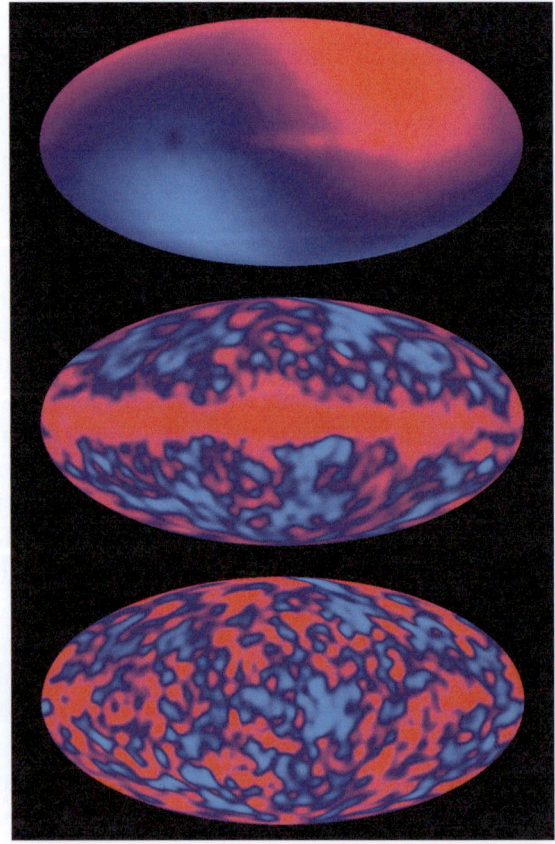

projection. If you also calculate out the Milky Way and other strong light sources, you get the image at the very bottom.; this now shows the smallest temperature differences ($10^{-5} \times 2{,}7$ K) of the background radiation.

The successor mission to COBE was *WMAP* (Wilkinson Microwave Anisotropy Probe), which explored this radiation from 2001 to 2010 and from 2009 to 2015 there was also the European probe *Planck*. These missions show the fluctuation of the background radiation with ever better resolution. Fig. 3.18 shows how the resolution of the fluctuations in the background radiation has increased over the course of the 3 satellite missions; a small section of the sky serves as an example. On the image that was obtained from COBE data (far left), hardly any details can be seen, with Planck many details are recognizable.

3.3 The Origin of Chemical Elements

The most common element in the universe is hydrogen, followed by helium and the rest of the chemical elements heavier than helium—which in astrophysics is referred to as *metal*—is almost negligible. But where do these elements come from?

3.3.1 The First Three Minutes—or the Primordial Nuclear Fusion

During the first three minutes of its evolutionary history, the universe was hot and dense enough to sustain a *nuclear fusion*. In nuclear fusion, lighter elements merge into heavier ones, with the end product being slightly lighter and the mass difference being radiated in the form of energy. Consider the fusion of hydrogen into helium. Four hydrogen nuclei make up one helium nucleus. This is 0.7% lighter than the four hydrogen nuclei combined, and this mass deficit is converted into energy. According to Einstein's theory of relativity, energy and mass are equivalent, also referred to as an equivalence.

As mentioned, only during the first three minutes of the universe's development were the conditions met for about a quarter of the existing hydrogen (more precisely protons, because there were no neutral hydrogen atoms at the high temperatures) to be fused into helium. Therefore, the composition of the universe is 3/4 hydrogen and about 1/4 helium. The result of this primordial nucleosynthesis was, in addition to ^{4}He, also the helium isotope ^{3}He as well as deuterium.

- ^{4}He consists of 2 protons and two neutrons in the nucleus,
- ^{3}He consists of 2 protons and one neutron in the nucleus,
- Deuterium is an isotope of hydrogen and consists of one proton and one neutron in the nucleus

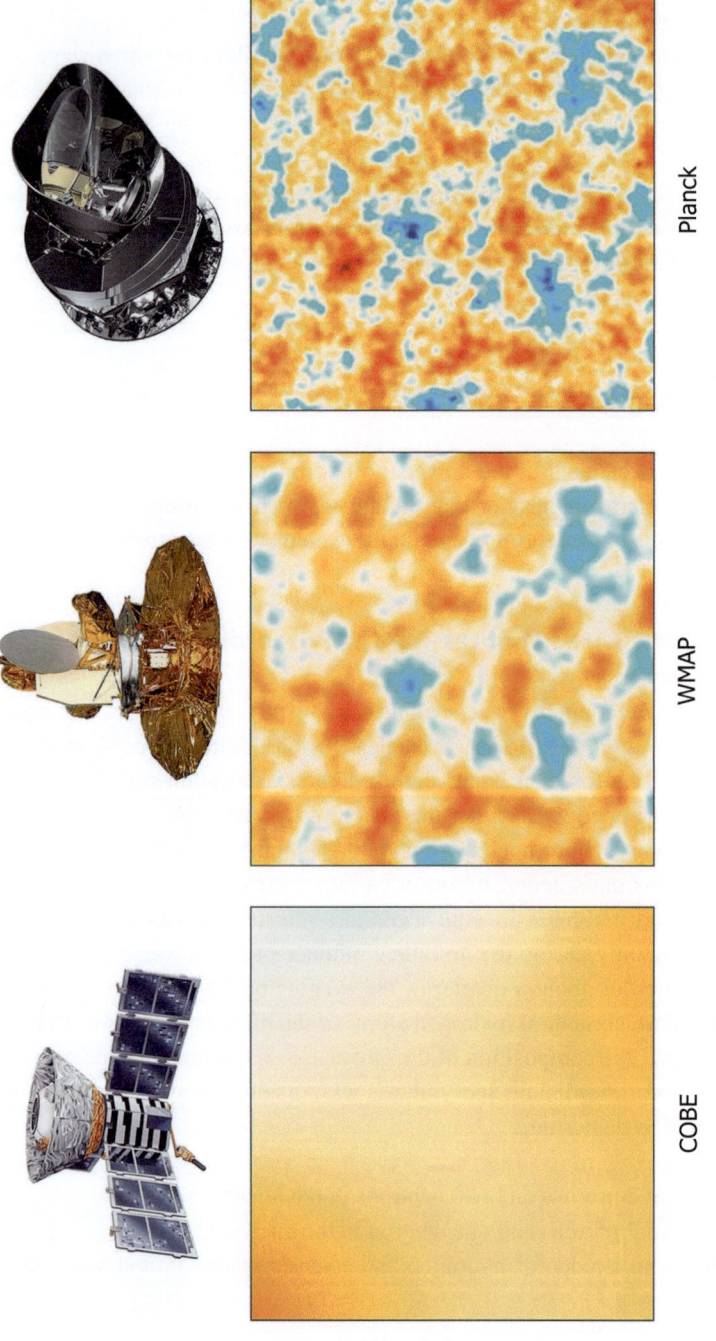

Fig. 3.18 Comparison of the recordings of a small sky field by COBE, WMAP and Planck

3.3 The Origin of Chemical Elements

As an example, consider the fusion of two deuterium nuclei into the helium isotope ^{3}He. As can be seen in Fig. 3.19, two deuterium nuclei react and form a helium nucleus, emitting high-energy photons (γ-quanta).

In the Fig. 3.20, the origins of the elements of the periodic table are color-coded. These are elements that occur in our solar system, but with minimal changes, this is applicable everywhere in the cosmos. The elements hydrogen and helium, highlighted in blue, are primordial, meaning they were formed in the first few minutes of the universe. A very small proportion of lithium as well. Elements such as boron, lithium, and beryllium are also produced by the splitting of atomic nuclei by high-energy particles of cosmic radiation. The elements highlighted in green originate from exploding massive stars at the end of their development. Other elements come from colliding neutron stars or dying low-mass stars, etc. We will deal with the development of stars in more detail. However, it is important to note here that all elements heavier than helium were produced inside stars or during their explosion.

▶ All elements heavier than helium were formed inside stars or at the end of their development. So, we are essentially made of stardust!

Fig. 3.19 Formation of the helium $He - 3$ nucleus by fusion of two deuterium atomic nuclei

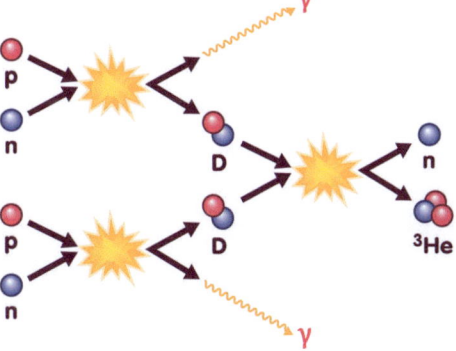

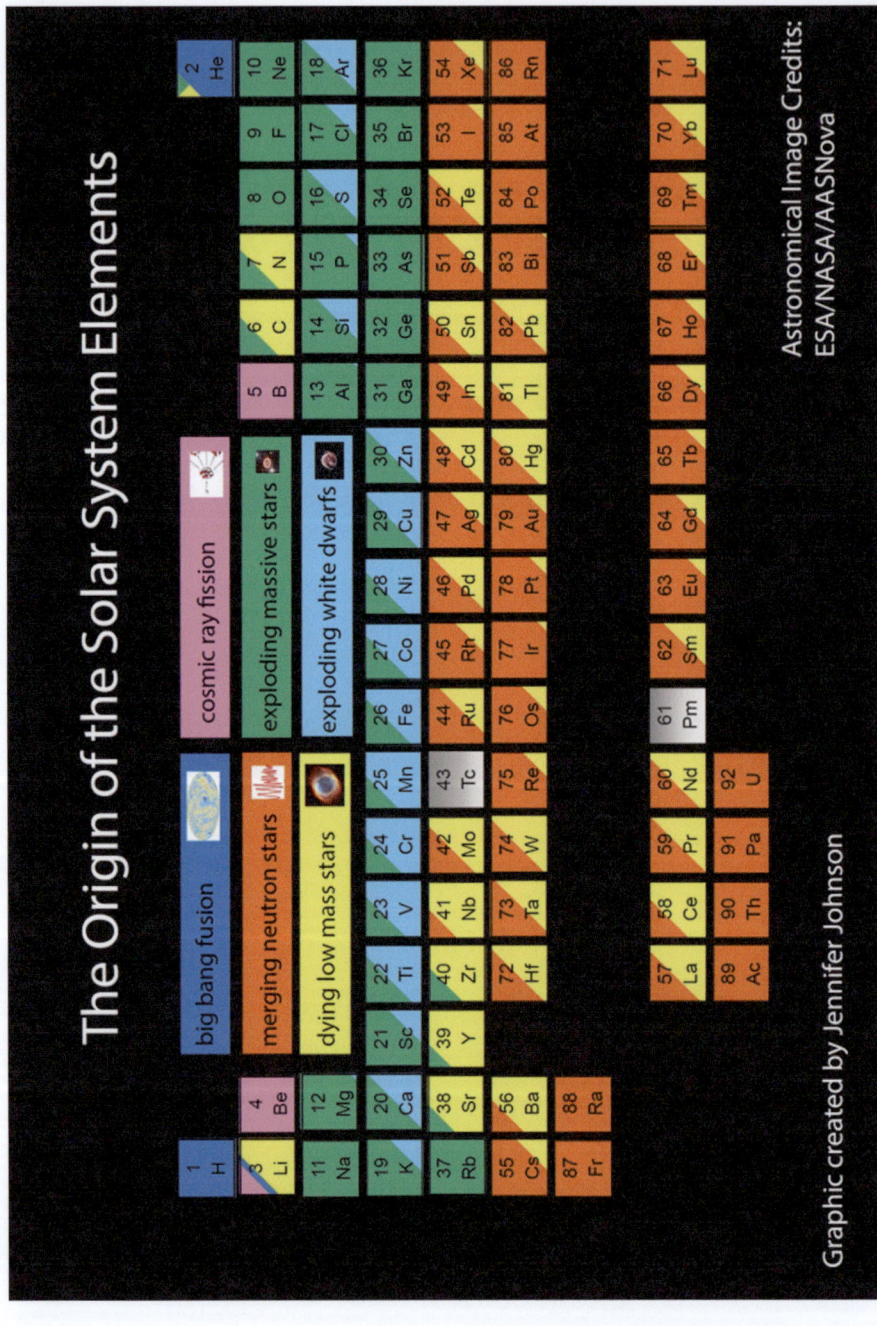

Fig. 3.20 The origin of the elements in the solar system

The Radio Sky

4

Our Earth's atmosphere is permeable not only in the optical range but also in the radio range. In this chapter, we discuss radio astronomy and show what exciting objects can be observed with it: from the formation of the first stars to the shadows of giant black holes.

4.1 What is Observed in the Radio Range

4.1.1 Radar Signals Explore Planets

Visible light is only a very small area of the entire electromagnetic spectrum. The human eye is sensitive in a wavelength range of about 400 nm to 700 nm; short waves around 400 nm we see as blue, long waves around 600 nm we see as red. Beyond the visible range of red, the range of infrared radiation follows, which we perceive as heat, but cannot see. The *microwaves* are in a wavelength range of a few mm to about 30 cm, which corresponds to a frequency of 0.3 THz to 1 GHz. The prefix T in the frequency unit *Hertz* (1 Hz = 1 oscillation per second) means 10^{12}. In Fig. 4.1, the range of microwaves is given in comparison to other wavelengths.

Due to the high frequencies, a high information density can be transmitted and microwaves are hardly affected by haze, clouds or rain. Thus, large amounts of data can be transmitted in a straight line using microwaves. Over short distances, *Bluetooth* uses microwaves. In a microwave oven, microwaves in the range of 12 cm, which is 2.45 GHz, are used. These microwaves penetrate the food a few centimeters and the water molecules inside the food absorb these microwaves and are set into oscillations. Therefore, the food heats up from the inside out. Microwaves are also used as navigation aids for airplanes and ships in the form of *Radar*; Radar is an abbreviation for Radio Detecting and Ranging. How does radar work? A radar device emits short electromagnetic

© The Author(s), under exclusive license to Springer-Verlag GmbH, DE, part of Springer Nature 2025
A. Hanslmeier, *New Windows into the Universe*,
https://doi.org/10.1007/978-3-662-71372-3_4

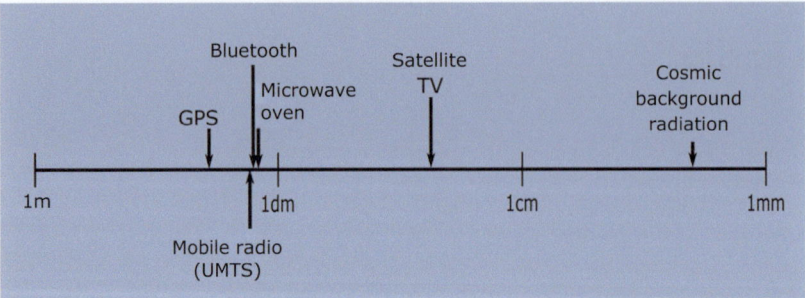

Fig. 4.1 The range of microwaves with adjacent radio range

pulses and receives the echoes. From the runtime from the transmission to the reception pulse, information about the nature of the reflector is obtained.

Using radar, for example, our neighboring planet *Venus* could be seen through the dense atmosphere. The US probe *Magellan* scanned with radar the surface of the planet from 1989 to 1994. The device was named VOIR (Venus Orbiting Imaging Radar). The probe orbited the planet and scanned its surface with radar beams. The map of Venus obtained from these observations is shown in Fig. 4.2 shown (Fig. 4.3).

Another application of radar technology is in *distance measurements* within the solar system. A radar beam is sent towards a planet and the time until the beam reflected off the planet's surface can be registered again is measured. The strength of the radar echo signal is proportional to $1/d^4$, where d is the distance to the object. Therefore, a high transmission power and sensitive receiving devices are very important (the transmitting and receiving antenna can indeed be different telescopes).

With the help of radar astronomy, it was also possible to obtain an image of *asteroids* as shown in Fig. 4.4 shown.

Radar observations also allow for Doppler shift measurements and, for example, the rotation of *Mercury* could be determined. On *Mars,* attempts are being made to investigate the ground to a depth of several meters using radar observations and to clarify whether and how much frozen water there is.

An example of a radar observatory is the NASA Goldstone Observatory located in California. The largest radio antenna there has a diameter of 70 m. With the observatory, up to 30 satellites can be observed or communicated with at the same time. It is also used for radar astronomy. The following projects are primarily carried out at the Goldstone Observatory:

- Radar mapping of objects in the solar system (planets, asteroids, comets).
- Observation of *quasars,* which are quasi-stellar objects; these initially appear star-shaped, but detailed examination shows that they are located at extremely great distances and therefore cannot be stars, but rather cores of active galaxies.

4.1 What is Observed in the Radio Range

Fig. 4.2 A radar map of the planet Venus; 98% of the Venus surface was mapped. NASA/Magellan

- Other radio sources in the sky, e.g., pulsars; these are very rapidly rotating *neutron stars;* the light from these objects is bundled and whenever a beam of radiation hits us, we see the object light up; the rotation period of these end stages of the development of massive stars is less than one second.
- Neutrino interactions on the lunar surface. These are produced by particles of cosmic radiation with extremely high energy (UHE particles, Ultra High Energy). These particles have an energy of up to 10^{20} eV. These energies cannot even be achieved at the CERN particle accelerator. Where do these particles come from? One explanation is the so-called *Fermi acceleration mechanism*: in this process, a particle's energy

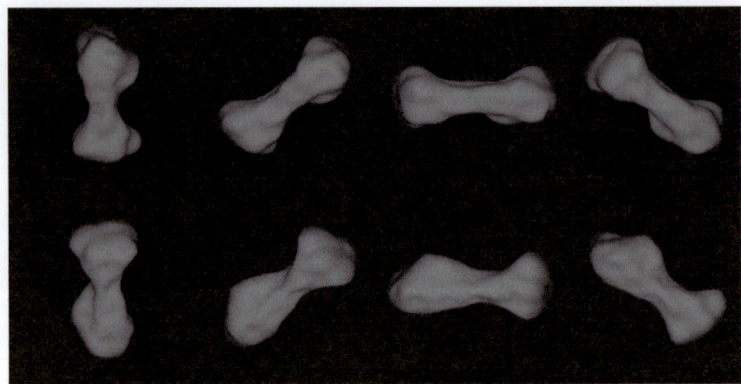

Fig. 4.3 A radar image of the asteroid Cleopatra. JPL

Fig. 4.4 The NASA Goldstone Observatory. NASA

increases through repeated interaction with shock waves. Another explanation could be the decay of massive particles that were formed during the Big Bang. About 20% of these particles come from *Centaurus A,* a bright radio galaxy 11 million light years away. In any case, observing these neutrinos requires extremely sensitive sensors.

Fig. 4.5 K.G. Jansky, the discoverer of radio emission from the universe

4.1.2 Discovery of Radio Emission from the Universe

K.G. Jansky (1905–1950, Fig. 4.5) is considered the founder of radio astronomy. He discovered a source of unidentifiable radio waves. In 1932, he was able to show that the origin of these radio waves was in the constellation Sagittarius. We know today that the center of our Milky Way is located there. If one could see radio waves with the naked eye, then the galactic center would be the brightest object in the sky.

The center of the Milky Way photographed in the visible range is shown in Fig. 4.6. The location of the radio source Sagittarius A^* is marked with a red circle. In the optical range, due to the strong absorption of dark dust clouds, almost no stars can be seen.

4.1.3 How a Radio Telescope Works

In principle, radio telescopes work similarly to optical telescopes. You need a surface that collects and focuses the radio waves; the detector is located at the focal point and after appropriate amplification, the radio signal can be displayed, for example, as a *false color image,* assigning different colors to individual radio frequencies.

The receiver here is the *radio antenna,* which usually takes the form of a parabolic mirror. As we have seen, the roughness of a mirror must not exceed $1/10\lambda$, where λ is the wavelength at which one observes. For the observation of radio waves with a wavelength of 1 m, an antenna that can be a grid is sufficient, with the mesh size having to be less

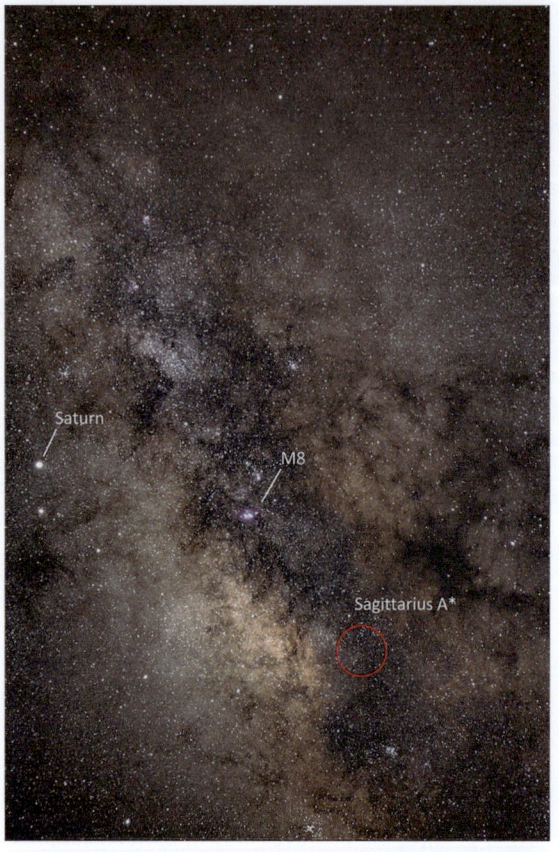

Fig. 4.6 Location of the radio source Sagittarius A*, the center of the Milky Way. At the time of the recording, the planet Saturn was nearby. C. Bergner

than 1/10 m. If one observes in the microwave range, the antenna must have a smooth surface. The antenna points to a point in the sky; to scan the sky, one either uses the rotation of the earth, or the antenna must be movable. A problem in radio astronomy is the resolution of the telescopes. Compared to optical wavelengths, radio waves are longer by a factor of 10^6 and more, accordingly, the *resolving power* deteriorates.

> A short calculation example: let's assume we have a 1-m telescope available. Then its resolving power in the visible range is about 0.1 arcsecond; how large is the resolving power in the radio range with a wavelength of 1 m and the same diameter?
>
> $$A = 206\,265 \frac{\lambda}{d} \tag{4.1}$$

for $\lambda = 1$ m and $d = 1$ m follows: $A = 296\,265'' = 57{,}3$ degrees in the sky. With a 1-m radio telescope, one could therefore only locate the radio radiation from objects in the sky up to about 60 degrees accurately. This is of course unsatisfactory.

So how can the resolving power of a radio telescope be improved? On the one hand, by building as large radio antennas as possible, on the other hand, by connecting several radio telescopes, which are at a distance d from each other, through *interferometry*. The resolving power then corresponds to that of a single telescope of size d. Two radio telescopes, for example, separated by 1000 km, then provide a resolution of about 0.2 arcseconds when observing in the meter range.

▶ To achieve good resolving power in the radio range, large antennas and the interconnection of several telescopes as far apart as possible are necessary.

4.1.4 How Radio Emission Occurs

Radio waves have a large wavelength and low energy. They are produced in different ways by astronomical objects.

- Molecular clouds,
- the 21-cm radio emission,
- Synchrotron radiation.

The space between the stars is not empty but contains atoms, gas, dust, molecules. These condense in some places into clouds. Molecular clouds mainly contain hydrogen molecules H_2, where two hydrogen atoms combine to form a molecule, but complex molecules can also form at the appropriate density, including organic compounds such as amino acids. In the mid-1960s, numerous compounds such as the hydroxyl radical (OH), cyan (CN), water (H_2O), ethanol, etc. were detected with the help of radio telescopes. Complex molecules can only form if they are appropriately shielded from the short-wave radiation of nearby stars. The radiation emitted by these molecules is in the millimeter wavelength range. Giant molecular clouds are found in the spiral arms of galaxies, and their mass is several tens of thousands to several hundred thousand solar masses.

Very dense and cold molecular clouds are the places where new stars are formed. In Fig. 4.7 you can see a molecular cloud in the *Carina Nebula*. The nebula itself has a diameter of 200 to 300 light years and it is about 8000 light years away from us, but can only be seen in the southern night sky. The extent of the molecular cloud is only 2 light years. The radio emission is caused by the rotation and vibrations of the molecules.

Fig. 4.7 Molecular cloud in the Carina Nebula. ESA/NASA

Synchrotron radiation occurs whenever charged particles are accelerated in a magnetic field. This is usually the case in astronomy when hot plasma moves in a magnetic field. Examples of synchrotron sources are *pulsars, radio galaxies* and *quasars*. Quasars are active cores of galaxies. It is assumed here that matter falls into a so-called supermassive black hole. This black hole can contain several million solar masses. Supermassive black holes are found in the centers of galaxies, including our Milky Way contains one such.

An important radio wavelength for observing the structure of our Milky Way is the so-called *21-cm line*. This radio emission is not caused by transitions of electrons from different energy levels in the atom but in the following way: Electron and proton (hydrogen nucleus) possess a spin, simply put a certain direction of rotation. Now there are two possibilities:

- Electron and proton rotate in the same direction, it is said that nuclear and electron spin are parallel;
- Electron and proton rotate in opposite directions; it is said that electron and nuclear spin are antiparallel.

Fig. 4.8 On the formation of the 21-cm line of the hydrogen atom

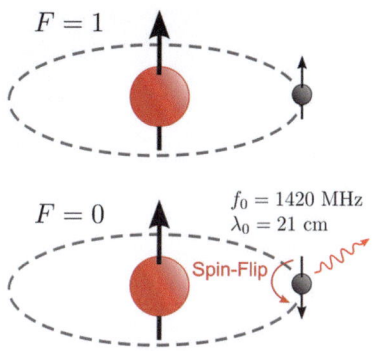

There is a very small energy difference between these two states and therefore, during the transition from parallel (referred to as F = 1) to antiparallel (referred to as F = 0), energy is released (an emission line is seen) or energy is absorbed (an absorption line is seen). This is illustrated in Fig. 4.8.

At which frequency can one observe the 21-cm line of hydrogen? The solution is quite simple:

$$c = \lambda \nu \qquad (4.2)$$

thus $\nu = c/\lambda = 3 \times 10^8 / 0{,}21 = 1428$ MHz.

The great importance of this line for radio astronomy lies in the fact that it is hardly attenuated (absorbed) by the matter between the stars.

4.2 Black Holes

4.2.1 Black Holes and Star Development

In this chapter, we briefly discuss the development of stars. Stars are formed by condensing interstellar gas and dust clouds. For such a cloud, which consists of many 1000 solar masses, to contract, it must become gravitationally unstable. This means that the inwardly acting gravity must be greater than the outwardly acting pressure forces (these are given by the gas pressure). The contracting cloud divides into several smaller clouds, which can in turn divide again. From these much smaller and less massive clouds, stars of different masses are then formed. From this scenario, it also follows that stars always form in groups, referred to as *open star clusters*. The term "open" indicates that the members of such a cluster do not stay together for all time, but that these clusters

Fig. 4.9 The Pleiades, also known as the Seven Sisters, in the constellation Taurus

dissolve. Our sun also originated in such a star cluster, but since these dissolve within a few 100 million years, it is very difficult to determine which stars were formed simultaneously with the sun 4.6 billion years ago. In young open clusters, remnants of gas and dust are still found, which are made to glow by young hot stars.

In the constellation Taurus, there is a well-known open star cluster, the *Pleiades (Seven Sisters)* (Fig. 4.9).

This star cluster is about 20 to 50 million years old and 400 light years away from us. Around the bright young stars, you can see glowing gas nebulae.

It is fascinating to observe how this star cluster (which is inexplicably sometimes confused with the constellation Ursa Minor) becomes noticeable in the night sky after midnight from late summer. The eye of the constellation Taurus, the orange glowing star *Aldebaran,* is also located in an open star cluster, the *Hyades;* however, Aldebaran is not a member of it (Fig. 4.10).

The stars initially shine due to the *gravitational energy* released during their contraction. In this process, the interior of the stars heats up and as soon as the temperature reaches several million degrees, the *nuclear fusion* ignites. Essentially, hydrogen burns to helium, causing stars to shine. Now, as they say in astrophysics, the star is in *hydrostatic equilibrium.* This is also the most boring phase in the development of a star - they hardly change and shine constantly. Only when the hydrogen supply in the core is exhausted, a new phase begins, nuclear fusion now takes place in a *shell* around the core and the star

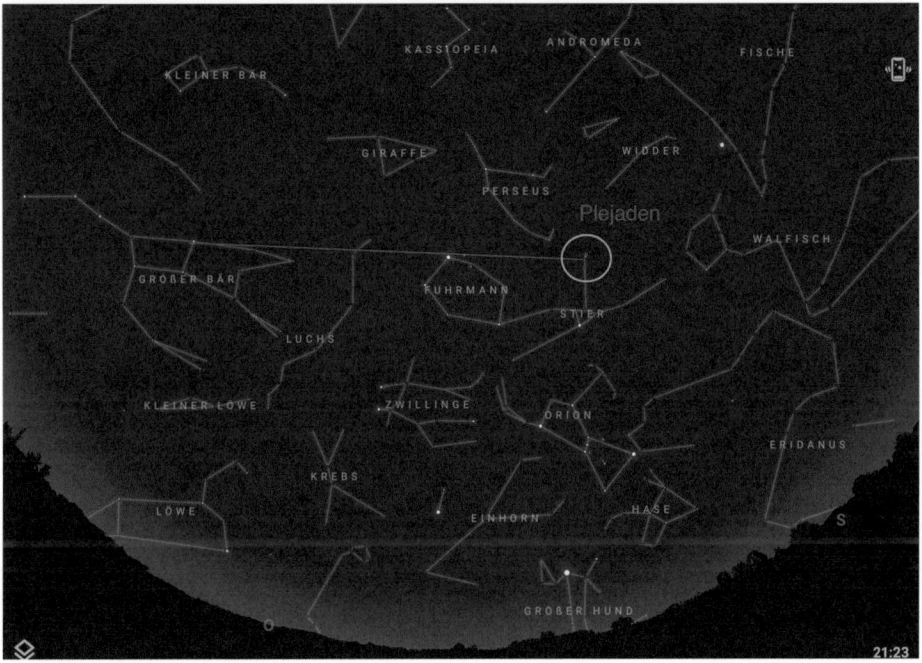

Fig. 4.10 In late summer, after midnight, you can see the Pleiades rising in the east, announcing the approaching winter in the starry sky. (Uni Oldenburg)

expands, it becomes a *Red Giant.* The core, which consists only of helium, contracts and gravitational energy is released again, i.e., the core heats up until the helium "ignites", i.e., the fusion of helium to other elements. In this phase, there are two energy sources inside the star: a shell spreading outwards, where hydrogen is still being fused to helium, and a core in which helium is fused to carbon. As a result, stars become unstable. Now the game starts all over again. Eventually, there is no more helium, we have a carbon core, helium and hydrogen burn in a shell around a contracting carbon core. As soon as its temperature is high enough, carbon burning occurs, and in this way all elements up to iron are created (^{56}Fe). Now it is important to note that there is a limit: the *Chandrasekhar Mass.* It is about 1.4 solar masses.

- Stars up to a mass of 1.4 solar masses end as *white dwarfs,* here the pressure of the so-called *degenerate electrons*[1] provides the counterpressure to prevent a complete contraction. Strictly speaking, this mass only applies to the iron core, but since the majority of a star's mass is concentrated in its core, we usually speak of the star mass

---

[1] Due to the Pauli principle, electrons cannot be packed arbitrarily close together.

and not the mass of the iron core. White dwarfs are very small, having approximately the diameter of the Earth (10,000 km).
- If the mass of a star is greater than 1.4 solar masses, the pressure of the degenerate electrons is no longer sufficient, they combine with the protons to form neutrons and the pressure of the *degenerate neutrons* can prevent a complete collapse of the star. However, the resulting *neutron star* is extremely densely packed and has a size of only a few 10 km! Neutron stars rotate very quickly. This is a consequence of the conservation of angular momentum. Imagine a figure skater preparing to spin. When she pulls in her previously extended arms, she spins much faster. The neutron stars, which are only about 10 km in size, have contracted from a star with a diameter of more than a million km!

  In addition, neutron stars have a very strong magnetic field in which charged particles are accelerated, and so the light is bundled and emitted like a lighthouse. We then observe a *pulsar,* a star that rotates around its axis in less than 1 s and emits radiation flashes, which can be observed in the form of regular light pulses. When the first pulsars were discovered in the sky in the 1960s, people thought they were receiving Morse code from extraterrestrials, and the objects were referred to as LGM, Little Green Men.
- At 3-4 solar masses, there is nothing at the end of a star's development that could stop the collapse, the star collapses completely and becomes a *black hole.* The gravitational field is so strong that not even light can escape. Black holes therefore appear completely black. However, nearby matter can be sucked into the black hole, creating friction and thus radiation (mainly in the X-ray range). The observation of X-ray sources in the vicinity of black holes is therefore an indication of the existence of a black hole.

The development of stars is summarized schematically in Fig. 4.11.

In Fig. 4.12 you can see the well-known Ring Nebula, M57. This is a so-called planetary nebula. When these objects were first observed with smaller telescopes, it was thought that they were small planetary disks, hence the name. In reality, these nebulae are created by the ejection of the outer shells of a star during its development into a white dwarf.

Note: the development into a neutron star or black hole is associated with a *supernova explosion*. The exploding star becomes as bright as an entire galaxy for a few weeks. Since this occurs exactly when the mass of the iron core exceeds 1.4 solar masses, all supernovae are equally bright, and the distance of the galaxy in which they light up follows from comparison with the apparent brightness.

4.2 Black Holes

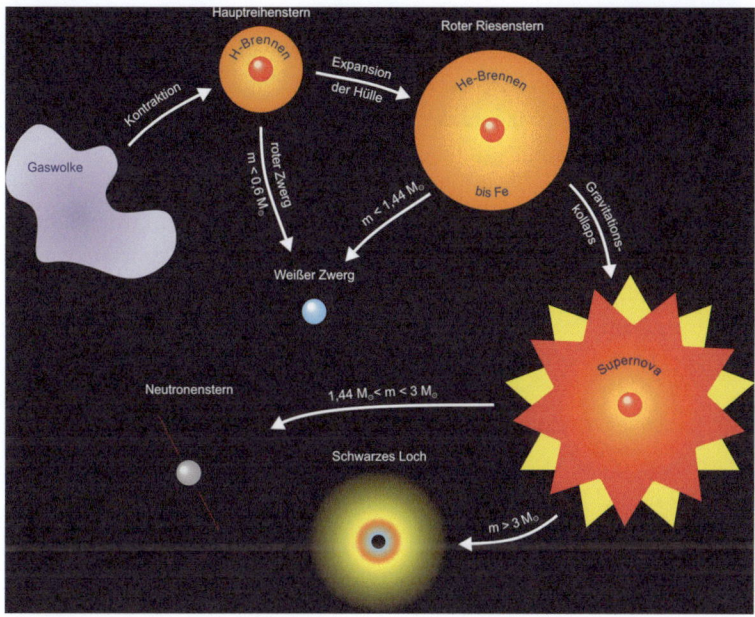

Fig. 4.11 Development of stars. Own work, GPL, https://commons.wikimedia.org/w/index.php?curid=1366291

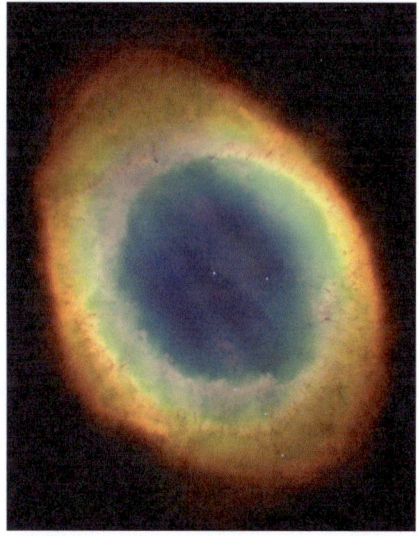

Fig. 4.12 A planetary nebula around a white dwarf located in the center. NASA/Hubble Telescope

Let's calculate how bright a supernova (a) would appear at a distance of 10 pc (b) in the Andromeda galaxy as seen from us. The absolute brightness of a supernova averages about −18 magnitude classes; remember, the absolute brightness corresponds to the apparent brightness at a distance of 10 pc; at this distance, a supernova would shine much brighter than the full moon in the sky. The extreme short-wave radiation produced by the explosion would also destroy the Earth's atmosphere and at least largely annihilate life on Earth. Don't worry, the next potential candidates for a supernova are more than 700 light years away from us, so 200 pc. Now let's calculate the apparent brightness of a supernova in the Andromeda galaxy: d = 650 kpc (distance of the galaxy), M = −18 and we use:

$$m - M = 5 \log d - 5 \tag{4.3}$$

and get from it: $m = 6{,}0$ so a supernova lighting up in the Andromeda galaxy would just be visible to the naked eye under extremely good conditions.

▶   The mass of a star determines its ultimate fate: stars below 1.4 solar masses become white dwarfs, stars between 1.4 and 3-4 solar masses become neutron stars, and more massive stars become black holes.

Let's estimate the so-called *Schwarzschild radius*: the escape velocity to leave the gravitational field of a star of mass M and radius R is:

$$v = \sqrt{\frac{2GM}{R}} \tag{4.4}$$

Now we set $v = c$ and calculate the corresponding Schwarzschild radius R_s

$$R_s = \frac{2GM}{c^2} \tag{4.5}$$

From the *Schwarzschild radius*, there is no escape from a black hole, one falls inexorably into the singularity. In principle, all objects we know could be transformed into a black hole, one just needs to bring them to the Schwarzschild radius:

- Sun: the Schwarzschild radius for the sun is 3 km,
- Earth: the Schwarzschild radius for the Earth is 1 cm. If we were to compress our Earth to 1 cm, it would be a black hole.

The *lifespan* of a star depends on its mass. Our sun has a life expectancy of about 9 billion years, the last few hundred million years as a red giant and then practically infinitely long as a white dwarf. Stars with less than one solar mass evolve even slower, stars with more than one solar mass faster.

4.2.2 Supermassive Black Holes

We have already met the two *Keck telescopes*. Each of the twin telescopes has a mirror diameter of 10 m. When operating the telescopes as an interferometer, a very high spatial resolution is achieved. With these telescopes, the center of the Milky Way has been studied. This cannot be seen in visible light due to absorption by interstellar dust, but it can be seen in near infrared light. Since the telescopes are located at a height that allows observations in the near infrared, stars have actually been found around the galactic center and photographed over a longer period of time. As shown in Fig. 4.13, the stars near the galactic center move around an object. The images were taken over a period of about 10 years and from them, as can be seen in the figure, the path of the objects can be clearly derived. Note the size scale of 0.1 arcsecond given in the top left of the image. For all those interested in mathematics, we indicate how one can deduce the mass of the object from this movement and thus derive the presence of a supermassive black hole in the center of our Milky Way.

> To estimate the mass around which the objects depicted in Fig. 4.13 move, we need the exact form of the *Third Law of Kepler*:
>
> $$\frac{a^3}{T^2} = \frac{G}{4\pi^2}(M_1 + M_2) \qquad (4.6)$$
>
> a. major semi-axis of the orbit ellipse, T is the orbital period of the mass M_2, M_1 is the sought central mass around which the object moves; we can assume $M_1 \gg M_2, M_2 = 0$. If we consider the inner orbit ellipse (turquoise), then $a \sim 0{,}1''$ and $T \sim 10$ Jahre. The center of the galaxy is about 30,000 light-years away from us, which is about 9000 pc. This results in the major semi-axis of the orbit to be $9000 \times 0{,}1/206\,265 = 4{,}36 \times 10^{-3}$ pc. If we insert these (roughly estimated) values into the third law of Kepler, it follows that the mass of the central invisible object is several million solar masses.

▶ At the center of our galaxy is a supermassive black hole with several million solar masses

How big is this monster at the center of our Milky Way? We know, the Schwarzschild radius for the sun is 3 km, and the mass goes linearly into the formula for calculating the Schwarzschild radius. Therefore, the Schwarzschild radius (i.e., quasi the extent) of the

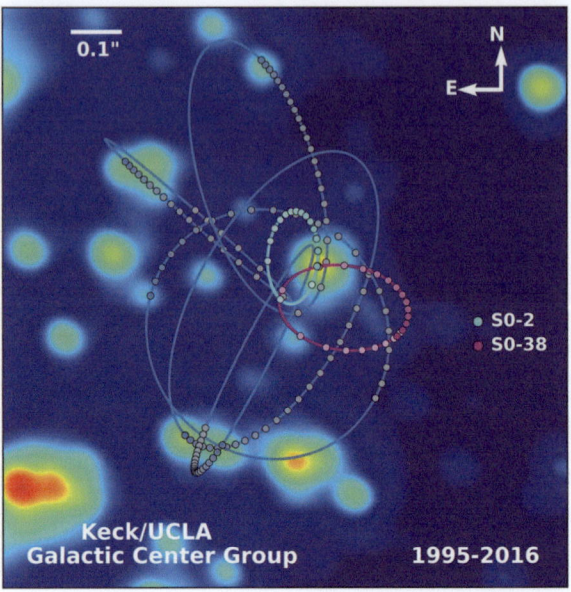

Fig. 4.13 Stacked images from the Keck Telescope of stars near the galactic center

supermassive black hole at the center of our Milky Way is several million km. For comparison: The Earth-Moon distance is about 400,000 km or the Mercury-Sun distance is between 46 and 69 million km. The supermassive black hole at the center of our Milky Way with several million solar masses would hardly be noticeable in our solar system unless matter falls into it.

4.2.3 Making the Monster Visible

We have seen how the existence of the supermassive black hole at the center of our Milky Way can be inferred through dynamics, that is, the movement of stars around it. However, the question is whether we can "see" this object, or at least something like its shadow, since the black hole itself is not visible. Radio astronomy offers a possibility for this. Radio telescopes distributed worldwide are linked together to observe the radio radiation emanating from Sagittarius A*. This project was given the resonant name *EHT, Event Horizon Telescope.* Event Horizon means event horizon, which is essentially the Schwarzschild radius. In Fig. 4.14, the radio telescopes involved in the EHT are shown. The radio signals measured from these observatories are combined through interferometry, achieving a spatial resolution almost equivalent to a radio telescope the size of the Earth.

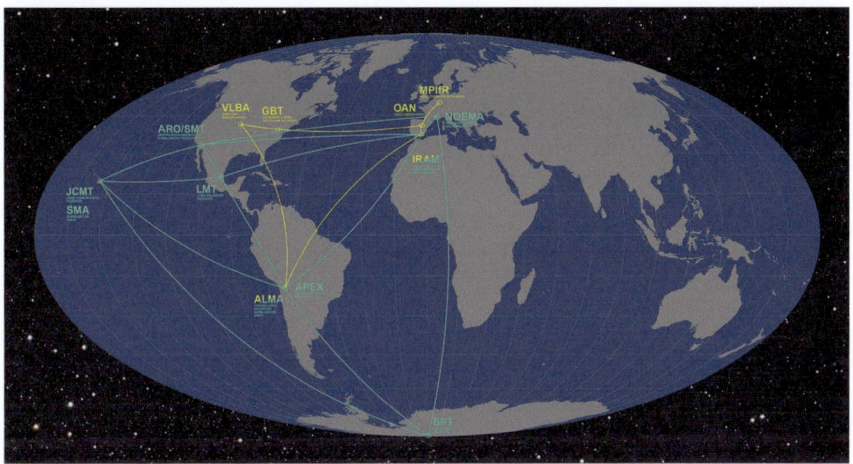

Fig. 4.14 The telescopes involved in the EHT project

- Resolution of the EHT: equivalent to a telescope with a diameter of 11,000 km
- Sensitivity (light gathering power or rather the ability to gather radio signals) corresponds to the sum of the areas of the individual telescopes.

To get an idea of the capability of this telescope, Fig. 4.15 shows that with its help, for example, an apple on the moon could be seen.

In the top right of the image, we see the first published capture showing the event horizon area of the black hole located in the core of the giant galaxy M87. This galaxy is approximately 55 million light-years away from us. The *accretion disk* (where matter flows into the black hole) distorted by the black hole shines brightly. The ring has a diameter of about 40 micro arcseconds and a width of less than 20 micro arcseconds[2]. The inner edge of the ring corresponds to the shadow of the black hole calculated from computer simulations. The bright spots are created by so-called relativistic beaming of light. The Schwarzschild radius itself corresponds to 4 to 7 micro arcseconds. The galaxy M 87 is shown in Fig. 4.16. One can recognize a *matter jet*, protruding from the inside. The length of this jet, which is connected to the supermassive black hole, is 5000 light-years. The galaxy is likely to contain about 10 times as many stars as our Milky Way. In principle, there should be two matter jets, but we only see the one pointing towards Earth. Here, the intensity of the light is increased by relativistic effects (beaming); the jet pointing away from Earth is too weak to be visible beaming (Fig. 4.17).

The corresponding image of the supermassive black hole (shadow) in the radio source Sgr A* at the center of our galaxy is shown in Fig. 4.18.

---

[2] One micro arcsecond = $1/1,000,000$.

Fig. 4.15 The resolving power of the EHT would make it possible to recognize an apple on the moon. The image of the supermassive black hole, or more precisely its shadow in the galaxy M81, is shown in the top right.

Although the two galaxies are completely different, M87 is a giant elliptical galaxy, our Milky Way is a spiral galaxy, the images look similar, however, the mass of the supermassive black hole at the center of M87 is 1000 times the mass of the supermassive black hole at the center of the galaxy. The EHT project involves 300 researchers from 80 institutes. It took 5 years to get the image 4.18.

4.2.4 ALMA

The acronym ALMA stands for Atacama Large Millimeter Array. The telescope, which consists of several (array) individual telescopes, is located on the Chajnantor plateau in the Chilean Atacama Desert and is operated by the European Southern Observatory along with other international partners. It can examine radiation of a wavelength of about 1 mm, thus covering the boundary area between infrared and X-ray radiation. The arrangement consists of 66 precision antennas, which are up to 16 km apart. Currently, this is the largest ground-based project in astronomy. Of these antennas, 54 have a diameter of 12 m and 12 antennas have a diameter of 7 m. The antennas can be aligned

4.2 Black Holes

Fig. 4.16 Image of the galaxy M 87 with matter jet. NASA/HST

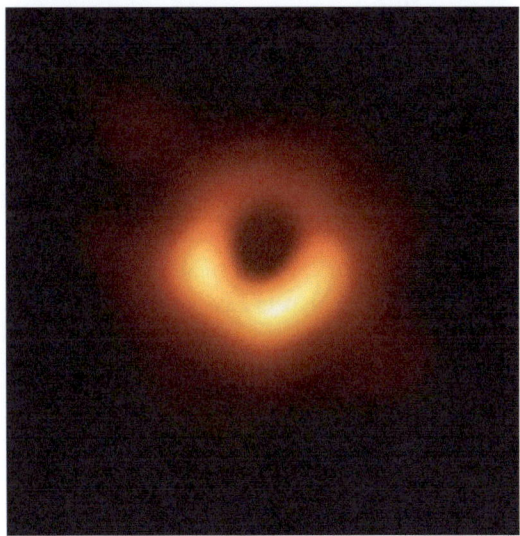

Fig. 4.17 Details of the shadow of the supermassive black hole at the center of the galaxy M87. EHT

with an accuracy of 0.6 arcseconds in the sky. This angle corresponds to the angle under which one could see a golf ball at a distance of 15 kmALMA.

What objects can be observed with ALMA? It is primarily about dense gas and dust regions; these are the places where new stars are formed. In visible light, nothing can be seen in these areas, they are impenetrable but in the millimeter and submillimeter range they are transparent. We thus see very cold objects. This wavelength range is also strongly absorbed when passing through the Earth's atmosphere, especially by the water

Fig. 4.18 Details of the shadow of the supermassive black hole at the center of our Milky Way. EHT

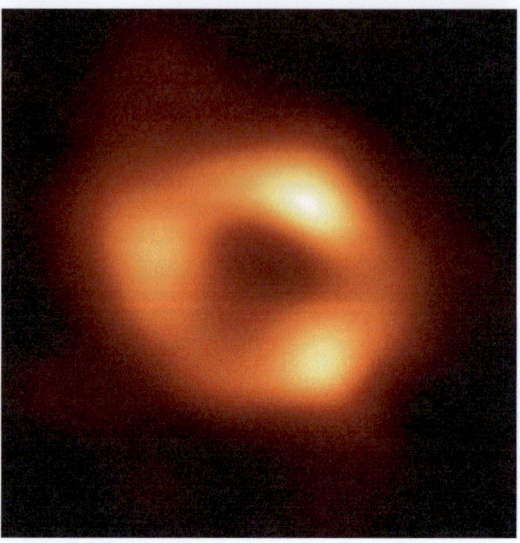

vapor contained in the air. The height of 5000 m, where the antennas are located, as well as the extremely dry location overall, greatly reduces the absorption by water vapor. The facility was completed in 2011. An illustration showing some of the antennas is given in Fig. 4.19.

In Fig. 4.20, you can see the formation of *protoplanetary disks* around a central star. The star *HL Tauri* is about 450 light-years away from us. Young stars are surrounded by a gas and dust shell, which flattens into a disk over time due to rotation. Clumps of matter form in this disk, which steadily grow into *planetesimals,* which are the precursors of the planets. Once planets have formed, they create gaps in the dust disk through gravitational resonances, and these gaps can be seen in the image. The finest recognizable structures have about a size that corresponds to the distance Sun-Jupiter. In visible light, the star HL Tauri cannot be seen.

Another example of ALMA observations are galaxies in the sub-millimeter range. Here, special galaxies have been studied that form new stars very productively. Such objects were discovered in the 1990s. Many stars and large amounts of dust are formed in these galaxies in a short time, so that many remain hidden behind dust clouds. Only with the help of submillimeter radio astronomy can these objects be seen (Fig. 4.21).

In Fig. 4.22, you can see an observation from the Orion Nebula. This nebula is about 1400 light-years away from us and can already be seen with the naked eye or with binoculars in the "sword" of Orion; in the submillimeter range, the spectral signatures of the molecule propionitrile can be recognized. The red curve shows ALMA measurements, the blue curve serves as a comparison and shows measurements in laboratories.

Molecules in space rotate and vibrate, and each molecule has a specific arrangement of rotational and vibrational states. Whenever a molecule transitions from one such state

Fig. 4.19 Some antennas of the ALMA array. ESA/ALMA

Fig. 4.20 The object HL Tauri with protoplanetary disks around it. Here one observes the formation of a planetary system. ESA/ALMA

to another, a certain amount of energy is either absorbed or emitted, often as radio waves with very specific wavelengths. Each molecule possesses a unique pattern of wavelengths that it emits or absorbs, and from this pattern, one can identify which molecule it is by comparing the measurement with laboratory data.

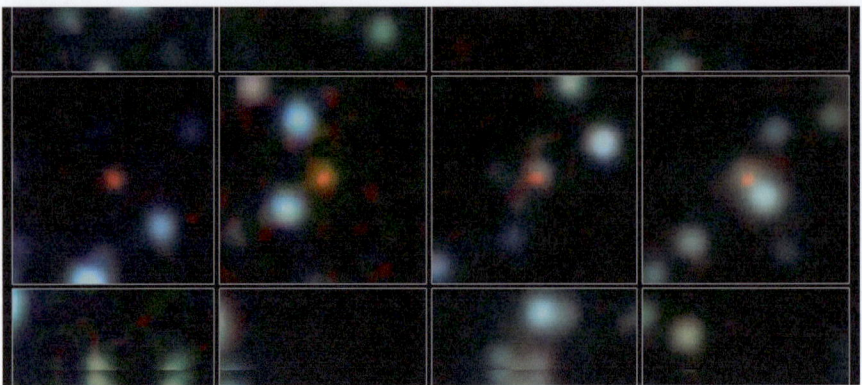

Fig. 4.21 Galaxies in the sub-mm range. ESA/ALMA

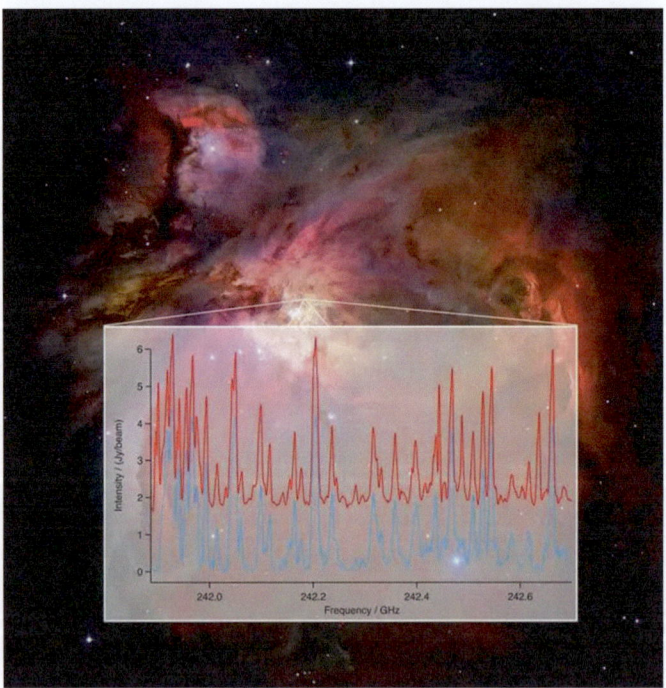

Fig. 4.22 Measurements of the molecule propionitrile. (C_3H_5N). ESA/ALMA. (FORTMAN, ET AL., NRAO/AUI/NSF, NASA)

Gravitational Waves—A New Window into the Cosmos

The term "theory of relativity" evokes awe and astonishment in many, coupled with the fear of never being able to understand it. Indeed, it is one of the most difficult chapters of modern physics, but its fundamentals can be understood through a few simple examples. Its results are both astonishing and baffling. For a deeper understanding of the topic, [3] is recommended.

5.1 General Theory of Relativity

5.1.1 Einstein and Relativity

We now come to a very complicated chapter of physics: the theory of relativity. Einstein has formulated two versions of this theory. The principle is always about how physical processes are represented in different reference systems. For example, if two events in one system are simultaneous (e.g., on a train), these events do not have to be simultaneous when viewed from another reference system. One might object that this does not correspond to our experience. The differences are most pronounced in the following example. Consider twins. Person A stays on Earth, Person B goes on a journey with a spaceship and is supposed to travel at a speed $v = 0{,}9c$, that is 90% of the speed of light. The person travels with the spaceship for 10 years (their speed reverses, but this is irrelevant), i.e., after 10 years, Person B returns to Earth. Person B says: "I have aged 10 years". However, the person who remained on Earth has aged more than 20 years! This is referred to as *time dilation*[1]. In very fast-moving objects, time passes more

---

[1] Dilation means expansion.

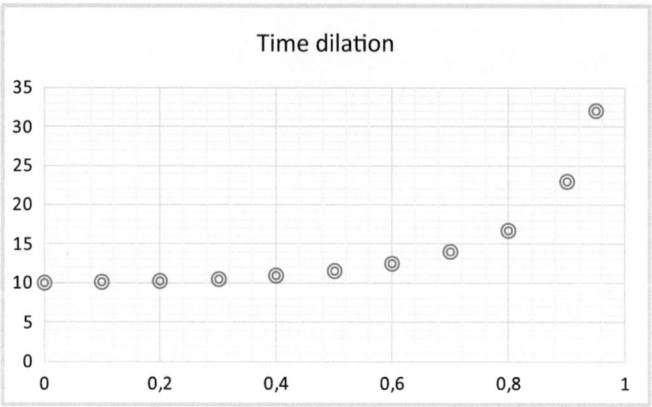

Fig. 5.1 The relativistic time dilation. If 10 years pass for a moving observer who is traveling at the speed plotted on the x-axis in units of the speed of light, the number of years that pass for the stationary observer is plotted on the y-axis.

slowly. This is shown in Fig. 5.1. The x-axis represents the speed in units of the speed of light (e.g., 0.5 means that one is traveling at 50% of the speed of light, or 150,000 km/s).

▶ Time dilation: time passes more slowly in a moving system.

So, do you live longer in a very fast-moving system than if you are at rest? The answer is ambiguous: yes and no. The observer in a moving system does not notice that time is passing more slowly, the difference is only noticeable when compared with the stationary observer (when the space travelers have returned to Earth from their mission).

We can easily calculate relativistic effects from the so-called *Lorentz transformation*. We only provide the final formula for time dilation here:

$$\Delta t' = \frac{\Delta t}{\sqrt{1 - \frac{v^2}{c^2}}} \tag{5.1}$$

v is the speed (e.g., of the spaceship), c is the speed of light, Δ always means an interval in physics, so Δt is the proper time that passes for the observer moving at speed v and $\Delta t'$ is the time that passes on Earth. Let's quickly make an example: suppose we have a spaceship moving at 99% of the speed of light, so $v/c = 0.99$. For the observer in the spaceship, 10 years should pass. Then it will be:

$$\Delta t' = \frac{10}{\sqrt{1 - 0.99^2}} = 71 \tag{5.2}$$

5.1 General Theory of Relativity

So while 10 years pass for the space-traveling individuals, 71 years have passed on the stationary Earth.

There is also a *relativistic mass increase*. The closer the speed of the body approaches the speed of light, the more its mass or inertia increases and eventually approaches infinity for $v = c$. Again, it should be noted that this is not noticeable at speeds that are small compared to the speed of light.

We calculate up to which speed the relativistic mass increase is less than 1%. The formula for the relativistic mass increase is:

$$m = \frac{m_0}{\sqrt{1 - \frac{v^2}{c^2}}} \tag{5.3}$$

where m_0 is the rest mass, that is, the mass that a body possesses when it is at rest. Now, $\Delta m = m - m_0 < 0{,}01 m_0$ should be, so

$$\Delta m < 0{,}01 m_0$$
$$m - m_0 < 0{,}01 m_0$$
$$\frac{m_0}{\sqrt{1 - \frac{v^2}{c^2}}} - m_0 < 0{,}01 m_0 \,|\, : m_0$$
$$\frac{1}{\sqrt{1 - \frac{v^2}{c^2}}} - 1 < 0{,}01 \,|\, + 1$$
$$\frac{1}{\sqrt{1 - \frac{v^2}{c^2}}} < 1{,}01 \,|\,^{-1}$$
$$\sqrt{1 - \frac{v^2}{c^2}} > \frac{1}{1{,}01} \,|\,^2$$
$$1 - \frac{v^2}{c^2} > \frac{1}{1{,}0201} \,|\, - 1$$
$$-\frac{v^2}{c^2} > \frac{1}{1{,}0201} - 1 \,|\, \times (-1)$$
$$\frac{v^2}{c^2} < \frac{0{,}0201}{1{,}0291} \,|\, \times c^2$$
$$v^2 < c^2 \frac{0{,}0201}{1{,}0202} \,|\, \sqrt{}$$
$$v < 0{,}14 c$$

> Therefore, for a relativistic mass increase of 1%, the object must move at 14% of the speed of light, which is 42,000 km/s, that is, once around the Earth in one second!

Let's calculate the relativistic mass increase for the Earth, which moves around the Sun at 30 km/s. This corresponds to $v/c = 30/300000 = 1/10000$.

Let's put this into the formula for mass increase:

$$m = \frac{m_0}{\sqrt{\left(1 - \frac{v^2}{c^2}\right)}} = \frac{m_0}{\sqrt{1 - 0{,}0001^2}} \tag{5.4}$$

we therefore obtain the following result: the mass of the Earth increases by a factor of 1,000,000.01 due to its speed of 30 km/s around the Sun. This is therefore a barely measurable very small effect. For this reason, the motion of the planets can also be determined using Newton's classical non-relativistic equations.

▶ Effects of the theory of relativity only occur at speeds that come into the range of the speed of light, roughly speaking, at speeds below 30,000 km/s they are hardly detectable.

An important statement of the theory of relativity is the existence of a maximum speed, the *speed of light*. We see this from the formulas given above. As soon as $v = c$, we have in the denominator of the equations for mass increase or time dilation $\sqrt{1-1} = 0$; this results in a division by zero, which is undefined. Only massless bodies (more precisely bodies with rest mass zero) can move at the speed of light.

▶ Masses cannot move at the speed of light

5.1.2 From Special to General Relativity

The examples we have dealt with so far are indeed very impressive and seem to contradict what we observe daily; but as mentioned, this is because the speeds in daily life are small compared to the speed of light. But there is a flaw: the equations only apply to reference systems that move against each other at a constant speed. When Einstein published this around 1905, he was aware of this limitation. But it took 10 years before he formulated the general theory of relativity, which is valid for arbitrarily moving systems, including accelerations. We remember: acceleration is a change in speed. Since speed is mathematically a vector (you drive a car at a certain speed in a certain direction; vectors have a magnitude and a direction), there are two possibilities for a change in speed:

5.1 General Theory of Relativity

- The magnitude changes, e.g., you accelerate from 0 to 100 km/h within 10 seconds.
- The direction changes, you drive into a curve, the earth moves on an elliptical orbit around the sun, etc..

Einstein recognized the *equivalence of inertial and gravitational mass;* that sounds complicated, but it's easy to understand. You probably know the feeling when you're in a quickly accelerating elevator. If the elevator goes up, you feel slightly pushed down, if it goes down (only when starting, that's an acceleration) you feel slightly lighter. The acceleration, i.e., the change in speed of the elevator, is supposed to be a_L. We are attracted by the earth with the earth's acceleration g. Our mass is m, our weight on earth: $G = mg$, then the following applies

- When driving upwards: our weight increases slightly $G = ma = mg + ma_L$
- When driving downwards: our weight decreases slightly: $G = ma = mg - ma_L$

In free fall, we are weightless. Astronauts orbiting the Earth are also weightless, they are essentially falling around the Earth. We speak of a *heavy mass* when it is attracted by a celestial body (e.g., Earth). We speak of an inertial mass when we mean the property of a mass to set itself in motion. And Einstein says that both masses are the same, so

$$m_{tr}a = m_s g \tag{5.5}$$

m_{tr} is the inertial mass, m_s the heavy mass, on which the Earth exerts a force with the acceleration g. This has an interesting consequence. Let's assume we are in a spaceship without windows. We step on the scale and find that we have a certain weight. But where does this weight come from? Einstein says there are two possibilities: Either the spaceship is on a planet that is attracting us, so we have a heavy mass, or we are somewhere in the universe and the spaceship is accelerating, then the scale shows the inertial mass.

▶ We cannot distinguish whether we have a weight due to the attraction of a planet or because our spaceship is moving with acceleration.

If you are still unclear about what inertial mass is, imagine a full coffee cup when a train accelerates or when it goes around a curve. In both cases, there is acceleration.

The general theory of relativity is a theory of *gravitation*. Einstein goes a step further here and describes physics in a four-dimensional *space-time*. Everything we measure in the universe takes place at a certain time in a certain place. A place in the universe can be described by 3 coordinates (x, y, z), time by one coordinate t. So, *events* in the universe are given by

$$(x, y, z, ct) \tag{5.6}$$

So, time is multiplied by the speed of light to obtain a unit of length. According to Einstein, gravitation is nothing more than a *curvature of space-time*. The Earth thus moves

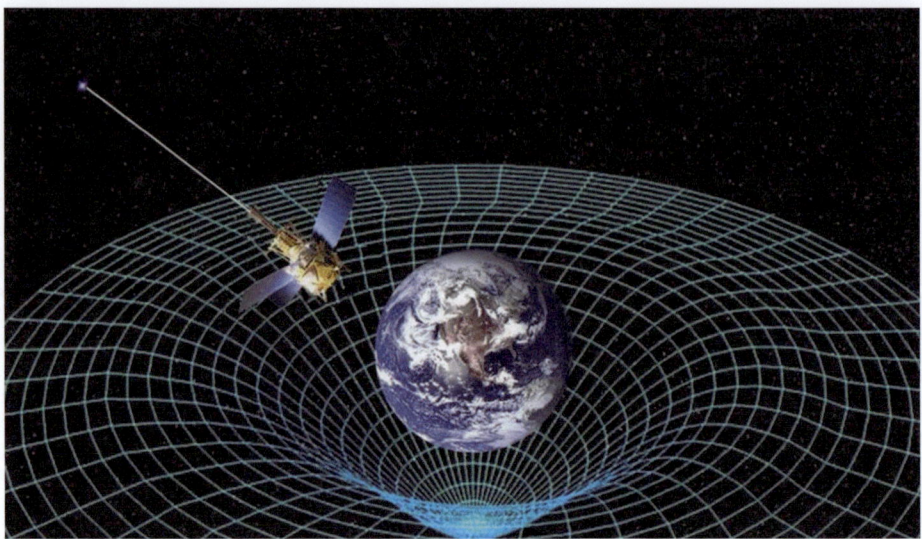

Fig. 5.2 The Earth creates a curvature of space-time around itself. A satellite moves along a geodesic line

around the Sun because the Sun, due to its large mass, curves space-time and the Earth moves on this curved space-time. In Fig. 5.1, the space-time curvature caused by the mass of the Earth is sketched. A satellite then moves along a line that is the shortest possible, this is called a geodesic (Fig. 5.2).

▶ General theory of relativity expressed in one sentence: Masses curve space-time

In a black hole, the curvature of space-time becomes infinite, therefore there is no escape within the event horizon. Naturally, many will now ask whether this can be measured at all. So, are there tests of the general theory of relativity? The answer was already hinted at briefly in Chapter 1. Light is deflected in a gravitational field, due to the curvature of space-time (Fig. 5.3). A beam of light follows the curved space-time, it cannot propagate in any other way, as there is nothing outside of space-time. We want to take a closer look at this first test of the general theory of relativity. Since the sun has a large mass, the curvature of space-time must be particularly large in its vicinity. Therefore, the idea arose to determine the position of stars during a total solar eclipse; the stars must be as close as possible to the sun's disk darkened by the moon and after some time their position was determined again, when the sun had moved far enough away (a few months later). By comparing the position measurements, the space-time curvature due to the sun was to be determined.

An initial expedition to measure this effect was undertaken shortly after Einstein's publication in the midst of the turmoil of the First World War to the Crimean Peninsula.

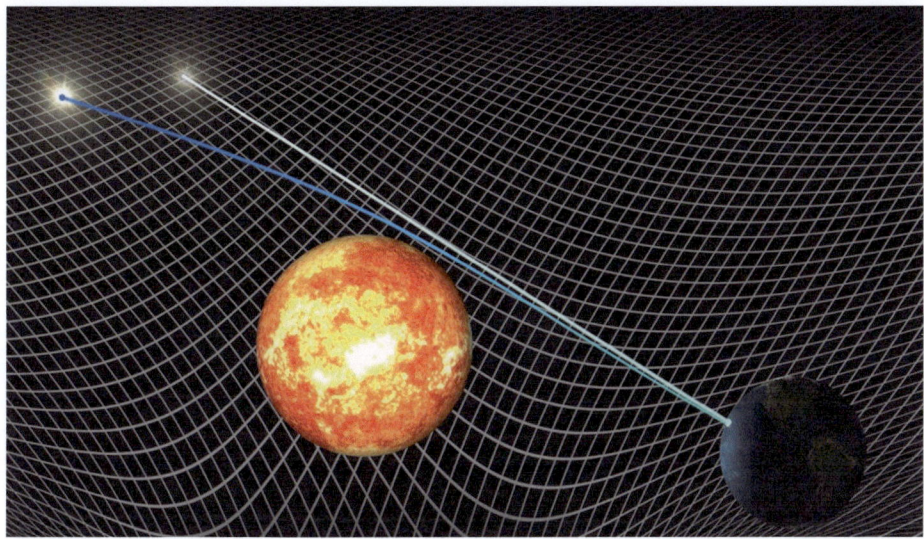

Fig. 5.3 The curvature of space-time by the sun leads to a deflection of light from nearby stars

However, the researchers were captured. Curiously, this was fortunate for Einstein, as the value of the shift due to space curvature he predicted was wrong, and so his theory would probably have been rejected if it had not been confirmed by the measurement. Einstein recognized the error and in the 1919 expedition, the *light deflection* could actually be measured, and the value agreed with that predicted by Einstein. Thus, Einstein was celebrated like a pop star, the only physicist to have been so. In Fig. 5.4 one of the photos from the observation of the solar eclipse on May 29, 1919 is shown. Some stars are marked that were measured precisely.

5.2 What are Gravitational Waves?

5.2.1 Waves

We speak of a wave when something spreads spatially with a certain period (i.e., a spreading oscillation). Mechanical waves are always bound to a specific medium. But there are also waves that can spread in a vacuum. These include electromagnetic waves (from X-rays to radio waves) and gravitational waves. Let's briefly summarize what we generally know about waves or what has been said about them in the previous chapters. Waves have the following properties:

- a wavelength: in the case of visible light, this is between 400 and 700 nm; radio waves range from cm to km.

Fig. 5.4 The total solar eclipse on May 29, 1919. The deviation of the stars due to the space-time curvature of the sun was measured for the first time

- Amplitude: this essentially determines the strength of a wave.
- a period P,
- a wave spreads; the propagation can, as in the case of *sound longitudinal,* occur parallel to the direction of propagation; but it can also occur perpendicular to the direction of propagation; we then speak of a *transverse wave;* the oscillation occurs perpendicular to the direction of propagation.

In Fig. 5.5, the propagation of a longitudinal wave and a transverse wave is sketched.

5.2.2 How Gravitational Waves Are Created

We remember how *electromagnetic waves* are created. In principle, always when *charges* are accelerated.

In Fig. 5.6, some examples of frequency ranges of electromagnetic waves are given.

Gravitational waves are created when *masses* are accelerated. Acceleration is always a change in speed. So, for example, when the Earth moves around the Sun, it constantly changes the direction of its speed. It thus emits gravitational waves. As a result, the system loses energy and the Earth will crash into the Sun due to radiation from gravitational waves. We are facing the same dilemma as atomic physics 100 years ago. If electrons orbit atomic nuclei, they must emit electromagnetic waves, lose energy, and crash into the nucleus. Quantum physics was able to show that electrons are stable on certain orbits.

5.2 What are Gravitational Waves?

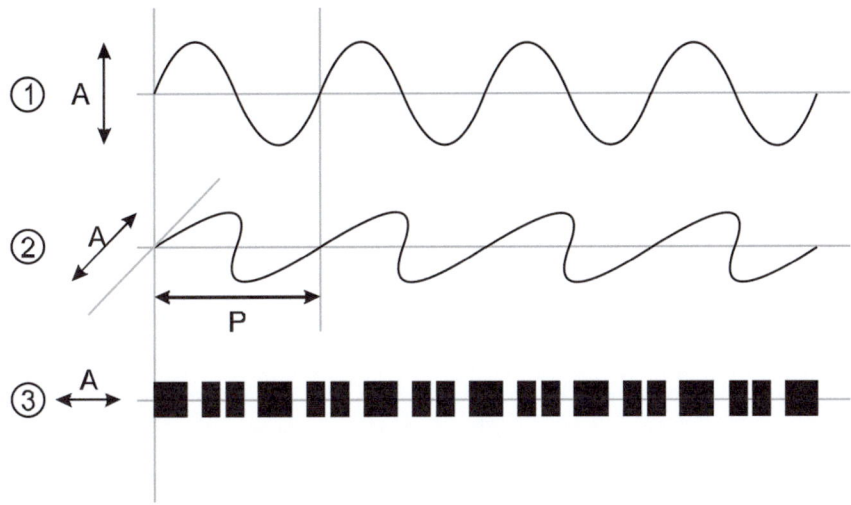

Fig. 5.5 Waves: **a** Longitudinal wave, **b** Transverse wave

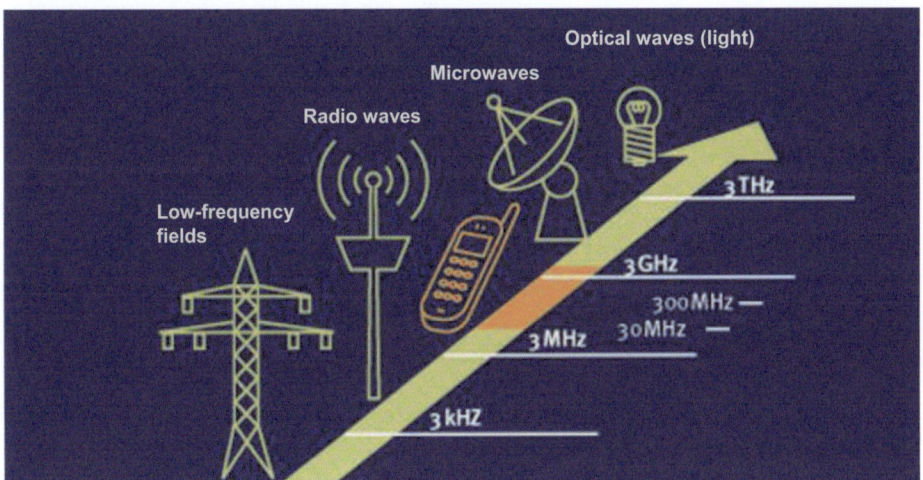

Fig. 5.6 Examples of frequency ranges of electromagnetic waves. According to SenUVK

Quantum physics does not help us here, but let's remember why we initially notice little of the relativistic effects in everyday life: the speeds are small compared to the speed of light. Therefore, one can expect that the energy loss resulting from the radiation of gravitational waves due to the Earth's orbit around the Sun will be very small.

A formula can be given for the power output of orbiting objects.

$$P = \frac{\left(\frac{r_{s,\odot}}{R}\right)^3 \left(\frac{r_{s,E}}{R}\right)^2 c^5}{5G} \qquad (5.7)$$

This applies to the Earth-Sun system. $r_{s,\odot}$ is the Schwarzschild radius for the Sun = 3000 m, $r_{s,E}$ is the Schwarzschild radius for the Earth, 0.01 m, $G = 6{,}67 \times 10^{-11}$ the gravitational constant and $c = 3 \times 10^8$ m the speed of light. R is the radius of the orbit, so 150×19^9 m. By substituting these values, we get a power radiation of 190 W. This formula applies to the idealized case of circular motion. The kinetic energy of the Earth's orbital motion isgt

$$E_k = \frac{1}{2}mv^2 \qquad (5.8)$$

where $m = 5{,}97 \times 10^{24}$ kg is the mass of the Earth and $v = 30000$ m/s is the orbital speed of the Earth around the Sun. We therefore obtain the following amount of kinetic energy: $E_k = 2{,}688 \times 10^{33}$ Ws. If you multiply the radiation power of about 190 W by the number of seconds in a year, the energy emitted due to gravitational waves in 10^{18} years is only one millionth of the Earth's orbital energy. So an unmeasurable effect!! Don't worry, we won't be plunging into the Sun anytime soon because the Earth radiates gravitational waves due to its orbital motion around the Sun.

▶ Moving objects emit gravitational waves, the radiation of the Earth due to its orbit is not measurable.

5.2.3 When Stars Collide

Now let's consider another case. Let's imagine two stars, for simplicity's sake, each with a solar mass. These are supposed to orbit each other at a distance of 1 million km. For math fans, we provide a small Python program that simplifies the calculation.

In Fig. 5.7 the simple Python program is given.

We enter the corresponding values into the program:

- $m_1 = 1.98e30$, so the mass of the first component is equal to one solar mass ($1{,}98 \times 10^{30}$ kg),
- $m_2 = 1.98e30$
- $R = 1e9$ the distance in meters.

5.2 What are Gravitational Waves?

```python
#!/usr/bin/env python3
# -*- coding: utf-8 -*-
"""
Created on Thu Jan 12 10:30:42 2023

@author: arnoldhanslmeier
"""

G=6.67e-11   # Gravitationskonstane
c=3e8        # Lichtegschwindigkeit
m_1=1.98e30  # erste Masse
m_2=5.97e24  # zweite Masse
r_s1=2*G*m_1/c**2  # Schwarzschildradius erste Masse
r_s2=2*G*m_2/c**2  # Schwarzschildradius zweite Masse
R=150e9      # Abstand der beiden Massen, Kreisbahn
P=(r_s1/R)**3*(r_s2/R)**2*c**5/(5*G) # Formel für Strahlungsleistung durch GW
print("Durch GW abgegebene Strahlungsleistung",P)

# Berechnung der kinetischen Energie
v_2= (G*m_1/R)**(0.5)  # Kreisbahngeschwindigkeit der zweiten Masse
e_k=0.5 *m_2*v_2**2# kinetische Energie (Kreisbahn)
print("Kinetische Energie der Bahnbewegung:", e_k)
xx=1e18
p1=P*3.1e7* xx # abgegebene Strahlunsleistung nach xx Jahren
print("Nach",xx,"Jahren abgegebene Strahlung durch GW",p1)
```

Fig. 5.7 Simple Python program for calculating the radiation of gravitational waves from two objects orbiting each other. In this case for the Earth-Sun system

We need to experiment with the value *xx* in the program to find out after how many years the amount of energy radiated by gravitational waves comes into the order of magnitude of the kinetic energy of the orbital motion. It turns out that this is already the case after one billion years.

If one assumes that the two components were only 100,000 km apart, then the value of the radiation through gravitational waves already reaches the value of the kinetic energy of the orbital motion after 100,000 years.

5.2.4 Collision of Two Neutron Stars

Of course, it is not realistic to assume that there are two normal stars each with a diameter of about 1 million km, separated only 100,000 km from each other, orbiting around the center of mass. However, this can very well be assumed for two neutron stars. The diameter of a neutron star is about 10 km. If two neutron stars approach each other, for example, to a distance of 100,000 km, they will collide with each other in about 100,000 years due to the radiation of gravitational waves.

The collision of two neutron stars is one of the most spectacular events in the universe. Extreme conditions arise during the collision. The temperatures rise up to 800 billion K. In this process, elements with heavy nuclei can also be formed (e.g., gold). In Fig. 5.8, an artistic representation of such an event is shown. But what is the result of this merger? Usually, the pressure of the degenerate neutrons can no longer withstand

Fig. 5.8 Two neutron stars collide and merge. An extremely rare event. (Graphic: Robin Dienel/ Carnegie Institution for Science)

the strong gravity, and a black hole is formed. However, it could also end differently, as a *Kilonova*. On May 22, 2020, a so-called *gamma-ray burst* was observed with the telescope of the Swift satellite. Such bursts are expected when two neutron stars merge into a black hole. But a few days later, a brightly glowing supernova was found, which should not actually be possible, because no radiation can escape from a black hole that is formed as the end product of a fusion.

▶   Gravitational waves occur as soon as masses are accelerated.

5.2.5 Properties and Detection of Gravitational Waves

First of all, it should be noted that while Einstein predicted the possibility of gravitational waves propagating at the speed of light, he was aware that it would be extremely difficult to detect them.

The universe and all objects within it can be described in terms of the general theory of relativity through a four-dimensional space-time continuum. In this conception, gravitational waves are nothing more than the propagation of a disturbance in this space-time continuum. The disturbance propagates at the speed of light. Detection is successful when the effect of this disturbance on a detector can be measured. The detector must be positioned in such a way that other vibrations that exist on the Earth's surface have no influence; thus, if small deflections of the detector are measured, these should have

5.2 What are Gravitational Waves?

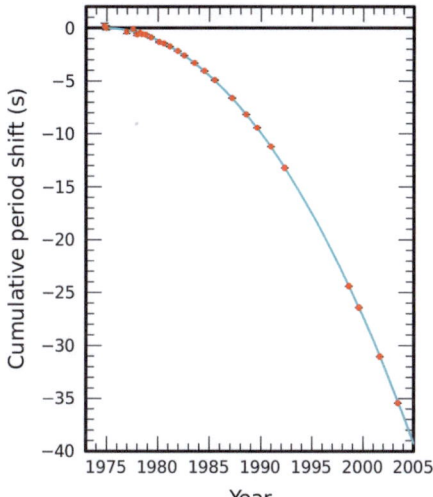

Fig. 5.9 Two pulsars approach each other due to radiation by gravitational waves

been caused exclusively by the propagating space-time disturbance. The expected deflections are also referred to as *strain* and are very small, approximately of the magnitude $h \times 10^{-18}$. h stands for the dimensions of the detector.

An initial indication of the existence of these waves was the observation of a binary pulsar PSR 1913+16, the orbital period of the two components changed as a result of radiation by gravitational waves. This can be measured with very high accuracy in pulsars. The components approach each other and the measured values agreed very well with the predicted values. For these measurements, R. Hulse and J. Taylor received the Nobel Prize in Physics in 1993. In Fig. 5.9, the decrease in the orbital period as a function of time is indicated.

In Fig. 5.10, the principle of a gravitational wave detector is illustrated. Two mirrors hung at points as far apart as possible (corresponding to the size h defined above) reflect light, which is brought to interference. The two mirrors hang on two perpendicular arms of equal length. If the distances between the arms change slightly due to an incoming gravitational wave, then the interference pattern also changes. Since the effect is very small, the light is reflected back and forth very often.

5.2.6 The Spectrum of Gravitational Waves

We have already mentioned some examples of when to expect the occurrence of gravitational waves, namely those where there is some prospect of successfully measuring them. Indirect evidence is possible with double pulsars, the first direct evidence was achieved with the gravitational wave detector LIGO. In Fig. 5.10 you can see which spectrum of gravitational waves can be observable and also some already completed or

Fig. 5.10 A gravitational wave detector

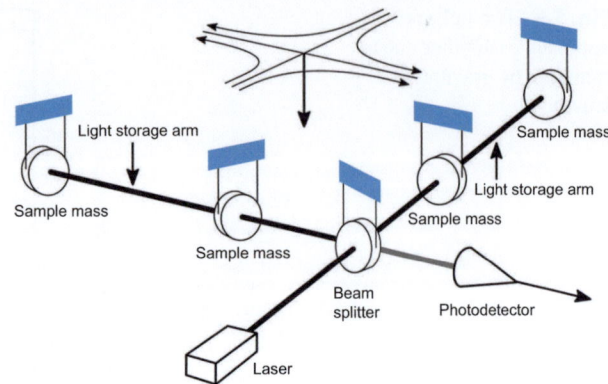

planned experiments for their measurement. With interferometers, gravitational waves from rotating neutron stars and compact binary stars in our Milky Way and beyond can be detected on Earth. The periods of the disturbances are extremely short, some 1/1000 s to one second. With the planned interferometers in space, for example, supermassive black holes in the centers of galaxies could also be studied. A completely different category of gravitational waves originates from the early days of the universe. These should be measurable through a polarization of the cosmic microwave background radiation (Fig. 5.11).

5.3 The First Direct Observation of Gravitational Waves

We have already discussed the principle of a gravitational wave detector. Laser beams that run back and forth between two perpendicular axes are brought to interference. The pattern changes as soon as a gravitational wave, coming from a cosmic event, compresses or stretches one of the two arms. To ensure that this is an event in the cosmos and not a disturbance on Earth, observations are made simultaneously at different observatories. This also has the advantage that the origin of the gravitational wave in the sky can be localized to observe the event with a telescope.

On September 14, 2015, the time had come. An event was measured with the LIGO experiment that lasted exactly 0.2 s. Sinusoidal waves with 10 to 15 cycles were observed, the amplitude of the wave increased to a maximum and then decreased again at a constant frequency. This observation matches exactly what is expected in a collision of two black holes (Fig. 5.12). The amplitude is proportional to the orbital speed of the black holes, which increases more and more until it almost reaches the speed of light. The frequency of the signal depends on the orbital frequency of the two objects, which also increases as they approach each other. The event happened at a distance of

5.3 The First Direct Observation … 123

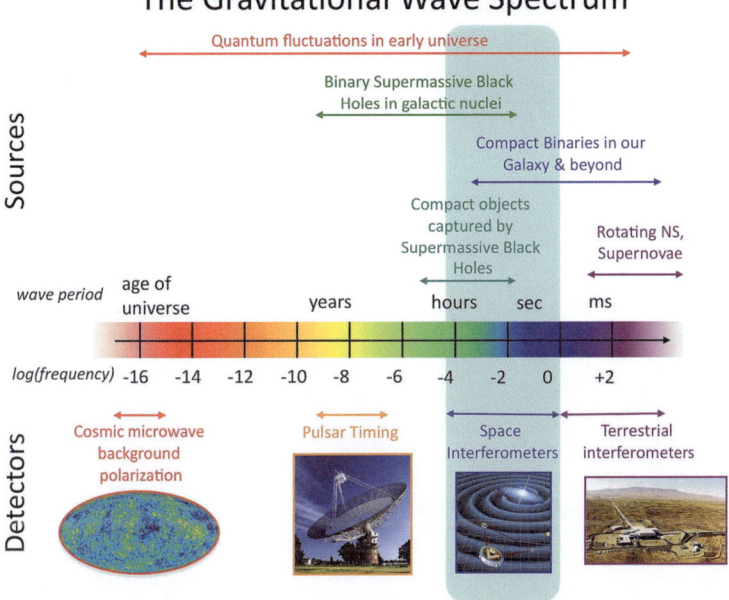

Fig. 5.11 The Spectrum of Gravitational Waves

Fig. 5.12 Two Black Holes Collide and Emit Gravitational Waves

410 megaparsecs, which is 1.3 billion light years. The two colliding black holes were very massive objects with 29 and 36 solar masses respectively. They merged into a black hole with 62 solar masses, 3 solar masses of energy were radiated in the form of gravitational waves. The event was given the designation GW150914.

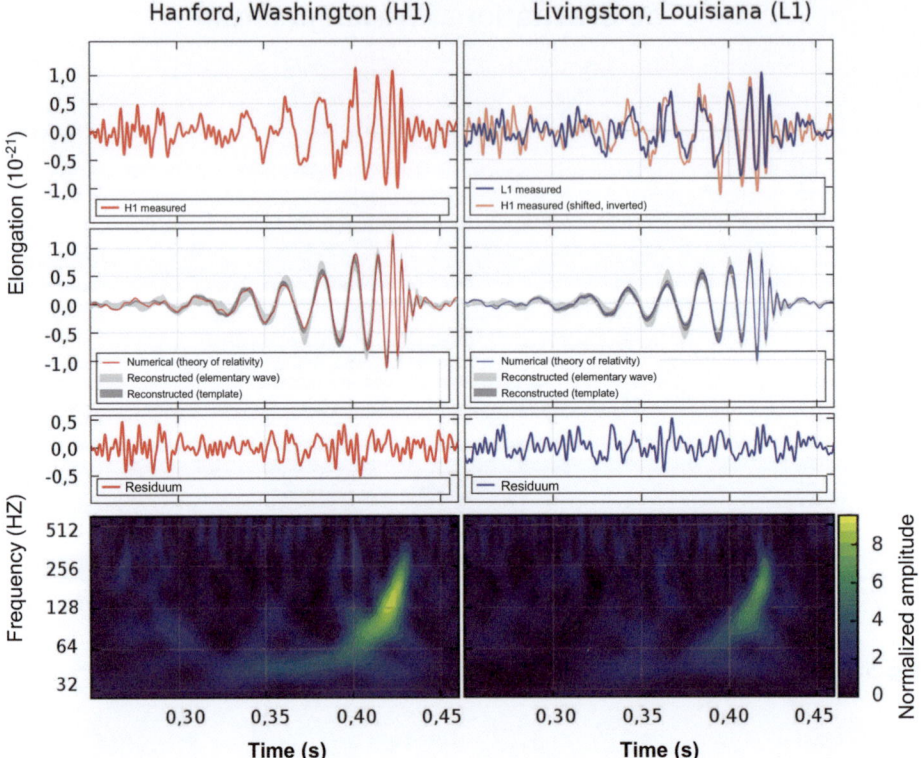

Fig. 5.13 Measurement of the gravitational waves emitted in the collision of two black holes. The measurements agree well with the model predictions

The event could be measured at two independent stations in Hanford (Washington) and Livingston (Louisiana). The measurement results of the two stations are shown in Fig. 5.13.

The measurements are offset by 7/1000 s. The reason for this is given in the insert for math fans.

> Why are the two signals from *GW150914* shifted by against each other? The light covers the following distance in this time:
>
> $$s = ct = 3 \times 10^8 \times 7 \times 10^{-3} = 21 \times 10^5 \, \text{m} \qquad (5.9)$$
>
> So that's 2100 km, which is roughly the distance between the two stations. Another proof of the accuracy of the measurements.

The result of the measurements of the event of September 14, 2015, was announced only after many tests and searches for possible error sources on February 11, 2016.

5.3.1 Further Measurements

On December 26, 2015, *GW151226* was measured. Here too, there was a merger of two black holes with 8 and 14 solar masses. On January 4, 2017, GW170104 was measured. The two black holes had 20 and 30 solar masses respectively and were 3 billion light-years away. A premiere was the measurement of the event *GW170814* on August 27, 2017. In addition to the two LIGO detectors, it could be measured with a third detector (VIRGO). This allowed an even more precise determination of the location in the sky.

On August 17, 2017, an event was then measured with the two LIGO and the VIRGO detector that lasted about 100 seconds. This is interpreted as gravitational waves released in the collision of two neutron stars . The masses were between 1.1 and 1.6 solar masses, the total mass was 2.7 solar masses. Only 1.7 s after the end of the gravitational signal, a gamma-ray burst was observed with the Gamma-ray Space telescope (Fig. 5.14) (designation *GRB 170817A*). The location could be identified with a telescope in the galaxy 130 million light-years away NGC 4993. In the infrared, UV and X-ray range, the afterglow could be observed, also referred to as a *Kilonova.*

In Fig. 5.15 you can see the measurements of the gamma-ray burst in different energy channels and below the gravitational wave signal averaged from the 3 observation stations (Fig. 5.16).

For the first time, it was thus possible to observe GW170817 observations in the field of gravitational waves and in various areas of electromagnetic waves. This is precisely the goal of all investigations.

5.3.2 Dark Matter and Gravitational Waves

The study of the movements of stars in our Milky Way and other galaxies reveals astonishing findings. Consider the movement of stars around the center of a galaxy. One would expect that the speed of rotation decreases the further away the objects are from the center. This is exactly what we know from the solar system: Mercury, the planet closest to the sun, orbits the sun in just 88 days, while Jupiter, which is about 10 times as far from the sun as Mercury, orbits the sun in about 12 years.

Let's now apply this to our Milky Way. Our sun takes about 220 million years to orbit the center. Stars that are even further from the center of the Milky Way than the sun should therefore move correspondingly slower. However, measurements clearly show that the orbital speed of stars far from the galactic center increases with increasing distance. Stars and objects far from the galactic center are thus moving too fast around it. There must therefore be a force that accelerates these stars; however, no necessary mass

Fig. 5.14 The Fermi Gamma-ray Space Telescope (Gamma-ray telescope), launched in 2008. NASA

is visible, hence the term "Dark Matter" was introduced. It makes up about five times the visible matter of a typical galaxy. Dark matter does not emit electromagnetic radiation, so it does not shine. But it acts through gravity and therefore the observation of gravitational waves offers the chance to learn something about the distribution of dark matter in the cosmos. More information about Dark Matter in [1] and [6].

5 Gravitational Waves—A New Window into the Cosmos

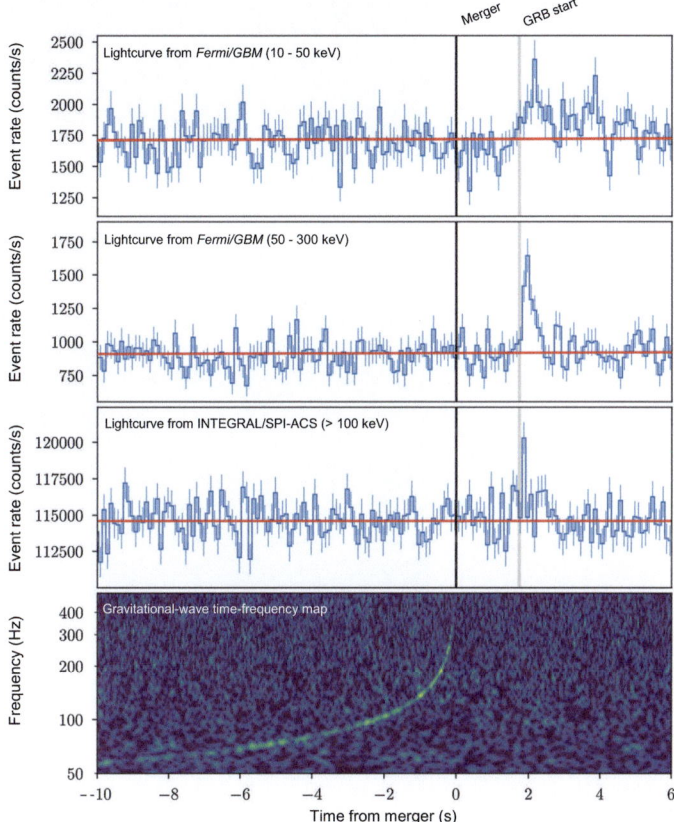

Fig. 5.15 The event GRB 170817A, in which gamma rays were detected by a satellite that originated from the merger of two neutron stars. The merger itself can be seen in the lower image by the typical gravitational wave signal

Fig. 5.16 The galaxy NGC 4993 with the brightening Kilonova (entered. Capture: Hubble Telescope. NASA

Neutrinos—Ghost Particles

Neutrinos are very special particles: they are electrically neutral, but pass through matter almost unhindered. Thus, they also offer a way to look into the interior of the sun and stars. We also discuss here some fascinating effects of the second major theory of modern physics, quantum theory, and the world of the smallest particles that everything consists of. For further reading, [12] and [4] are recommended.

6.1 What are Neutrinos

6.1.1 A Whole Zoo of Particles

When we look at the world of atoms, we encounter a multitude of particles. Some of them were initially thought to be elementary, meaning that these particles cannot be further broken down into smaller ones, but it turned out that there are only a few elementary particles that do not consist of even smaller units. Let's start with the structure of an atom. As mentioned in the previous chapters, atoms consist of a nucleus and electrons. The *nucleus* is made up of positively charged *protons* and uncharged *neutrons*. The nucleus is orbited by *electrons,* but there are also atomic nuclei without electrons. If the number of negatively charged electrons equals the number of positively charged protons in the nucleus, the atom is electrically neutral. *Ions* are charged atomic nuclei, here one or more electrons are missing and the atom is positively charged, or it has one or more extra electrons, then it is negatively charged. The neutrons contribute nothing to the charge. Atomic nuclei that differ only in the number of neutrons are called *isotopes*. The hydrogen atom has 3 isotopes:

1. Hydrogen atom with one proton (surrounded by an electron)
2. Hydrogen atom with one proton and one neutron in the nucleus (surrounded by an electron)
3. Hydrogen atom with one proton and two neutrons in the nucleus (surrounded by an electron)

So, in the case of hydrogen, there is always one proton in the nucleus. The isotope with an additional neutron is called *Deuterium,* the isotope with 2 neutrons in the nucleus is called *Tritium.* In all three cases, the atom is neutral when it is surrounded by an electron.

The next element in the periodic table of elements is *Helium.* This has 2 protons in the nucleus. In the isotope ^{3}He there is an additional neutron, in ^{4}He there are 2 neutrons. A ^{4}He$-$ nucleus is referred to as an $\alpha-$ particle.

Before we discuss the other particles, let's briefly look at some other properties of protons, neutrons, and electrons.

The mass of a *proton* is $1{,}6726 \times 10^{-27}$ kg. This corresponds to a rest energy of 938.272 MeV. According to the formula of Einstein

$$E = mc^2 \tag{6.1}$$

protons can be generated as soon as this energy is available. This is exactly what happened in the early universe, a few moments after the Big Bang; protons were formed when this energy became available. In a particle accelerator, one can reproduce the generation of particles at sufficiently high energies. The greater the mass of the particles, the more energy must be available for their generation. Protons are very small, the radius (more precisely the charge radius) is only 0.84 fm. 1 fm is a femtometer, a very small unit: $1\,fm = 10^{-15}$ m, that is $0{,}000\,000\,000\,001$ mm.

The *neutron* has a slightly larger mass than the proton: $1{,}6748 \times 10^{-27}$ kg, which corresponds to a rest mass of 938.565 MeV. Neutrons decay, if they are not bound in a nucleus, with a half-life of 10 minutes, the average lifespan is 15 minutes. This is referred to as *beta decay,* since electrons are produced, which were formerly also referred to as beta particles. An important law in atomic physics states that the electric charge must be conserved, no matter which processes one considers. Therefore, a neutron can only decay into a proton and an electron. However, precise measurements showed that something was missing, and this is where we come to the *neutrinos.* Antielectron neutrinos are produced in beta decay. As the name implies, neutrons are electrically neutral.

The representation of the decay of a neutron looks as follows:

$$n \rightarrow p + e^- + \bar{\nu}_e + 0{,}78\,\text{MeV} \tag{6.2}$$

n... Neutron, p... Proton, e^-... Electron and $\bar{\nu}_e$ Antielectron neutrino; it releases an energy of 0.78 MeV. Under normal conditions, it is difficult to measure the decay of a neutron, as it is quickly absorbed by other atomic nuclei. For practical applications, neutrons are therefore considered stable particles.

There is also the inverse β decay: this is important in the formation of the already discussed neutron stars. Protons and electrons combine at appropriate energy to form neutrons, and an electron neutrino is created.

$$p + e^- + 0{,}78\,\text{MeV} \rightarrow n + \nu_e \qquad (6.3)$$

Let's briefly consider the *electron*. The word electron comes from the Greek and means *amber*, as electricity was discovered through the friction of amber. Electrons form the shell around a nucleus. The mass of an electron is about 1/1800 of that of a proton or a neutron, more precisely $m_e = 9{,}109 \times 10^{-31}$ kg. The rest energy is 0.510 MeV. The charge of an electron is also referred to as *elementary charge*, it is negative and is often written as $-1e$, where the elementary charge $e = 1{,}6 \times 10^{-19}$ C amounts to. In metals, some of the electrons are freely movable, which is why they have electrical conductivity.

6.1.2 How elementary are protons and neutrons

Initially, it was thought that electrons, protons, and neutrons were true elementary particles, that is, indivisible. The idea that the world consists of the smallest particles dates back to the Greek philosopher *Democritus* (460–370), who first spoke of indivisible units (Greek atomos means indivisible) from which everything is composed: "Only apparently does a thing have a color, only apparently is it sweet or bitter, in reality there are only atoms in empty space." According to Democritus' conception, the soul should also consist of soul atoms, which are scattered upon a person's death and form a new soul with other soul atoms.

Only later were some of these philosophical considerations confirmed by experiments. However, our image has changed since then. In the modern *standard model of particle physics*, protons and neutrons consist of the so-called *quarks*. These are elementary, they cannot be divided into smaller units. Together with the electrons, the quarks are therefore the actual elementary particles. The quarks possess fractional charges and they come in 6 types, which is also referred to as "flavor" and for these flavors, the nice names up, down, charm, strange, top, and bottom have been introduced. These "flavors" (English flavors) of the quarks differ in terms of mass. Heavy quarks only occur in very short-lived hadrons. An overview of the quarks is given in Table 6.1 (Fig. 6.1).

The masses of the quarks are represented as a sphere in Fig. 6.2. On the left in the small picture, the comparison of the mass between electron and proton is shown.

Protons and neutrons consist of three quarks.

- Proton: 2 u and one d quark. This is briefly written as *uud*. The charge of the u-quark is $+2/3e$, of the d-quark $-1/3e$, thus we obtain for the charge of the proton $2/3e + 2/3e - 1/3e = +1e$.
- Neutron: 1 u, 2 d quarks. Thus, the charge of the neutron is $2/3e - 1/3e - 1/3e = 0$.

Table 6.1 Some properties of the quarks

Name	Symbol	Charge	Mass (MeV)
Down	d	−1/3 e	4.67
Up	u	+2/3 e	2.16
Strange	s	−1/3 e	93.4
Charm	c	+2/3 e	1270
Bottom	b	−1/3 e	4180
Top	t	+2/3 e	172760

Fig. 6.1 Democritus, the founder of the atomic theory of matter

The electron is an elementary particle, it cannot be further divided into smaller units.

▶ Quarks and electrons are elementary particles, protons and neutrons consist of 3 quarks.

But how can quarks of the same charge stick together (same charges are known to repel each other)? The *strong interaction* causes quarks of the same charge to form a proton or neutron, it is also said that the quarks are held together in the protons and neutrons by so-called *gluons* (glue in English means adhesive). And to make the whole thing even more confusing: Quarks possess, in addition to the classical charge (+ or -), another property, the so-called *color charge*. So there are red, green and blue quarks, where the colors are supposed to express this color charge property. And these 3 colors neutralize each other.

Fig. 6.2 Representation of the masses of the quarks as a sphere. On the left in the small picture, the comparison of the mass between electron and proton is shown

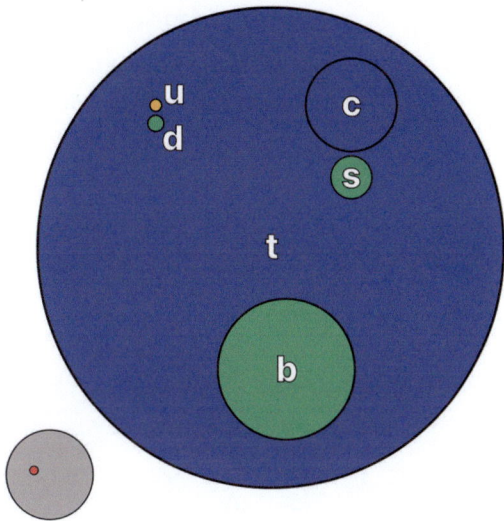

Fig. 6.3 The proton consists of 3 quarks uud

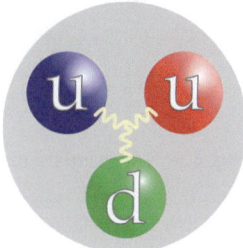

Fig. 6.4 The neutron consists of 3 quarks ddu

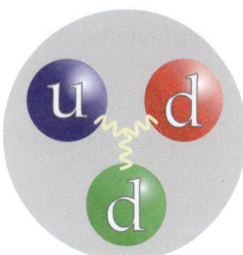

A sketch of how a proton or a neutron is constructed can be found in Figs. 6.3 or 6.4.

Leptons are elementary particles that do not undergo the *strong interaction*. The term lepton comes from Greek and means small, fine. There are three leptons: electron e^-, muon μ and tau particle τ, which differ in terms of mass (the τ−particle is heavy, so leptons are not just light particles). For each lepton, there is a neutrino, then there are the

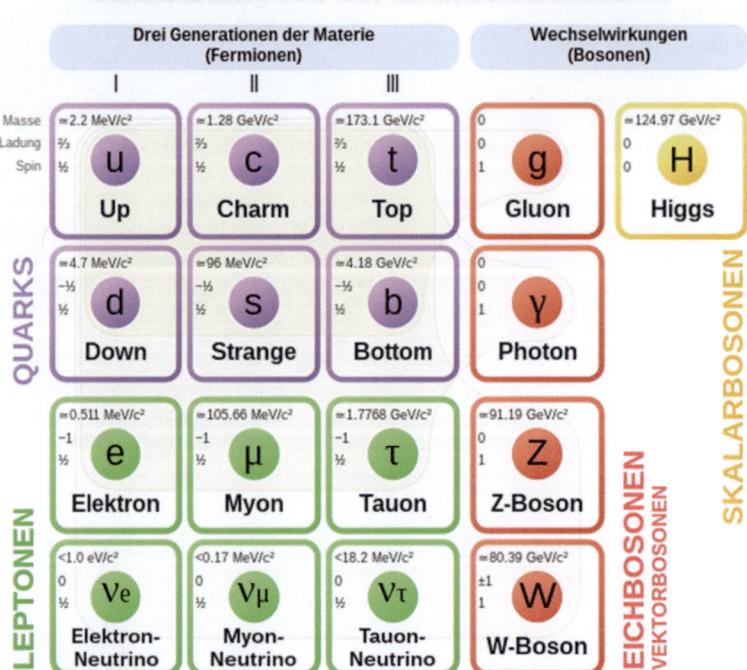

Fig. 6.5 The standard model of particle physics

discussed quarks and then the particles that are exchanged in an interaction, so-called *bosons*. The gluon is exchanged in the strong interaction, the photon in the electromagnetic and W and Z particles in the weak interaction.

In addition to the leptons, there are the hadrons: these include:

- Mesons: consist of 2 quarks
- Baryons: consist of 3 quarks

Protons and neutrons are therefore baryons. For each particle, there are antiparticles, which usually differ by the opposite sign of the charge.

Figure 6.5 shows the so-called standard model of particle physics. As you can see, the matter is quite complicated and it is quite correct to speak of a particle zoo. The Higgs boson belongs to the Higgs mechanism. It is supposed to be a Higgs field. All elementary particles get their mass through interaction with the Higgs field. And for each field, there are particles. We have already learned the term quantization in physics. Certain physical quantities such as charge, mass etc. do not appear in nature in arbitrary values (continuously) but only in integer multiples of a smallest unit (e.g. the elementary charge). This is also called *first quantization*. Quantum physics also teaches us that there is essentially no difference between particles and waves. Depending on the experiment, either the property of a particle or a wave is measured. The *second quantization* now says that a particle can be described as an excited state of a quantum field. The Higgs boson is thus the quantum mechanical excitation of the Higgs field. Gauge bosons, photons, and the gluons are massless. The photons mediate the electromagnetic interaction, the gluons the strong interaction; the weak interaction is mediated by the W and Z particles, which are not massless. They interact with Higgs field, which gives them their mass.

It should also be mentioned about quarks that they never occur as free particles but are always trapped in baryons, which is referred to as *confinement*. Only in the very early phase of the universe, at extremely high temperatures, there was a so-called *quark-gluon plasma*.

6.1.3 Neutrinos

We have mentioned these particles several times. They occur in beta decay, etc. Neutrinos are electrically neutral, so they have no charge. However, they are almost massless. The first ideas about neutrinos assumed that their rest mass, similar to that of photons, should be zero. There is a neutrino for each of the three leptons:

- ν_e Electron neutrino
- ν_μ Muon neutrino
- ν_τ Tau neutrino

For each of these three types of neutrinos, there is an antiparticle. This is indicated by a superscript bar, so for example, the anti-electron neutrino is denoted as $\bar{\nu}_e$. The mass of the neutrinos is extremely small, it is less than $1{,}5 \times 10^{-36}$ kg or the rest energy $< 0{,}8\,\text{eV}$. They basically only interact weakly with matter, which makes neutrinos so difficult to detect. Neutrinos can pass through matter almost unhindered. In Fig. 6.6 you can see a recording of an experiment in which a neutron was involved.

In Fig. 6.6 you can see the image of a reaction in a bubble chamber in which a neutrino was involved. A bubble chamber is usually filled with liquid hydrogen. Just before the start of the experiment, the pressure is reduced, the temperature of the hydrogen is above the boiling point, the incoming particles ionize the hydrogen atoms and serve as nuclei for bubbles that can be observed. There is usually also a magnetic field and the particles are deflected according to their charge (Fig. 6.7).

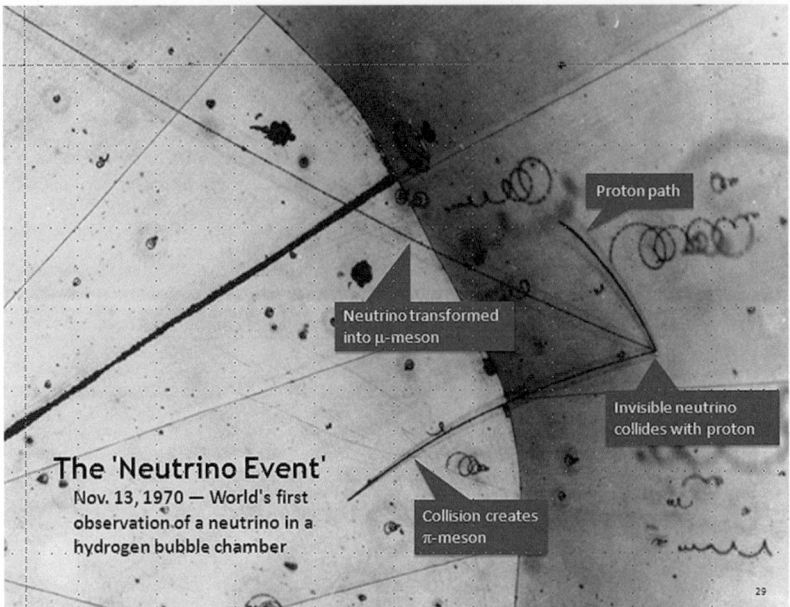

Fig. 6.6 With a bubble chamber, the tracks of particles can be made visible. The first recording of a neutrino in a bubble chamber filled with liquid hydrogen was made at the Argonne National Laboratory in 1970. A neutrino collides with a proton. The reaction took place on the right in the picture—where three tracks converge. The neutrino beam was obtained from decaying positively charged pions, which were produced by bombarding a beryllium target with the proton beam. Argonne Nat. Lab

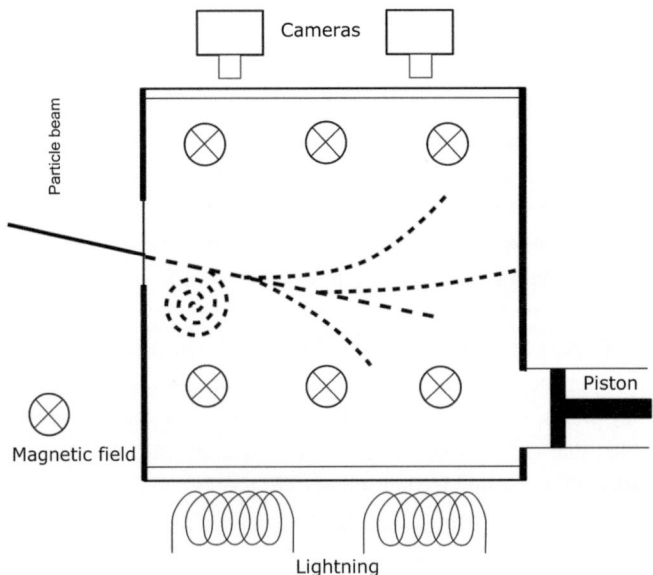

Fig. 6.7 Schematic setup of a bubble chamber with which particles can be made "visible"

6.2 Discovering Neutrinos

6.2.1 The First Neutrino Observatory

Neutrinos are extremely difficult to detect due to their very low interaction with matter. The first *neutrino observatory* was established in the 1960s under the direction of Raymond Davis. The detector is located 1478 m underground in the *Homestake Gold Mine* in South Dakota near Rapid City. The underground placement of the detector allows for the exclusion of other influences, e.g., from high-energy cosmic radiation. In Fig. 6.8, you can see a tank filled with 615 t of tetrachloroethylene, a cleaning fluid. The following reactions occur: some chlorine atoms, when hit by neutrinos coming from the sun, transform into argon atoms.

$$\nu + {}^{37}\text{Cl} \rightarrow {}^{37}\text{Ar} + e^- \qquad (6.4)$$

The low interaction of neutrinos with the atoms of the neutrino tank is demonstrated by the following number: On average, a reaction was detected every 2 days. It is truly like searching for the proverbial needle in a haystack to detect about 15 Ar atoms in the tank after a month (Fig. 6.8).

Fig. 6.8 A telescope of a special kind: deep underground, the first attempts were made to detect neutrinos originating from the sun

6.2.2 Additional Neutrino Telescopes

What can be measured with a neutrino telescope?

- Energy, direction, arrival time of the particles,
- Type of particle,
- Flux (how many particles per time and area),
- Correlation with astronomical observations.

The interaction of particles from the universe with the Earth's atmosphere is summarized in Fig. 6.9.

In this figure, we recognize the following types of interactions:

- The neutrinos (red) mostly pass through the entire Earth without interaction.
- Protons and nuclei react with hadrons and produce muons and neutrinos (through decays of so-called π particles).
- Muons lose energy through ionization and have a large range (e.g., several kilometers in water).

In recent decades, additional neutrino detectors or neutrino telescopes have been developed. All are located at great depths, as signals from muons generated by cosmic radiation as secondary particles in the Earth's atmosphere at the Earth's surface, exceed the neutrino signals. We only provide a very incomplete list here, as examples. In the *GALLEX Experiment* in Gran Sasso, Italy, neutrinos up to an energy of 233.2 keV are detected. The following reaction is used for their detection:

$$\nu_e + {}^{71}\text{Ga} \rightarrow {}^{71}\text{Ge} + e^- \tag{6.5}$$

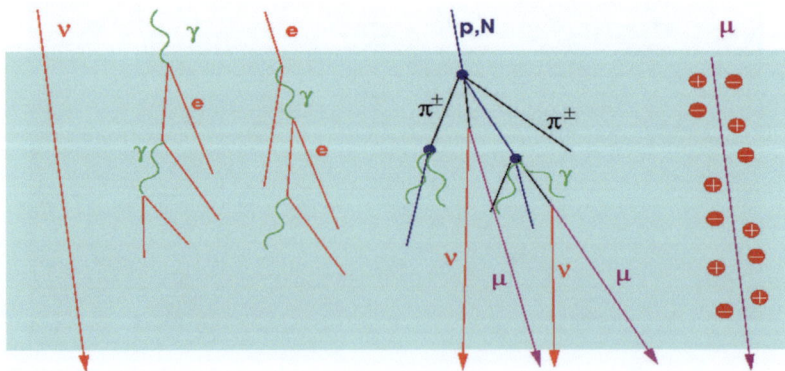

Fig. 6.9 Interaction of particles from the cosmos with particles of the Earth's atmosphere

Fig. 6.10 Super Kamiokande scintillation detectors for the detection of Cherenkov radiation flashes due to neutrino reactions. Kamiokande, Japan

Gallium is thus converted into germanium. *GNO,* in Gran Sasso and *SAGE* in Baksan, Russia, work similarly. In the already mentioned Homestake Experiment, the threshold for detection is 789 keV, so only neutrinos with higher energy can be detected. The *Kamiokande Detector* in Japan uses heavy water, and the reaction with fast neutrinos results in the emission of *Cherenkov radiation* which can be measured as light flashes (Fig. 6.10).

6.3 Where do Neutrinos from the Universe Come From

6.3.1 The Problem of Solar Neutrinos

Our sun is a star. As we have seen, stars produce their luminosity through nuclear fusion. Lighter atomic nuclei merge into heavier ones. Roughly speaking, inside the sun at a temperature of about 12 million degrees, four hydrogen nuclei merge into one helium nucleus. A hydrogen nucleus consists of one proton, while a helium nucleus consists of two protons and two neutrons. Therefore, two protons must be converted into two neutrons. This happens through the so-called inverse beta-decay.

$$p \rightarrow e^+ + \bar{\nu}_e + n \tag{6.6}$$

A proton here decays into an antielectron, positron, e^+ a neutron n and an antielectron neutrino $\bar{\nu}_e$. This exact neutrino was attempted to be detected in the first established neutrino observatory in South Dakota. We know the physical conditions inside the sun, especially temperature, pressure, and composition. Therefore, we can predict how many neutrinos should arrive on Earth, and since we know the cross-section of the neutrinos, we can also predict how many neutrino reactions should be detected per month. The number of neutrinos coming from the sun is enormous: per second, the sun produces about $1,8 \times 10^{38}$ neutrinos, each cm^2 of our bodies is penetrated by several trillion neutrinos from the sun, but as already emphasized, only extremely few of these neutrinos cause a reaction in the detector.

The sun emits about 10^{38} neutrinos per second. By how much does it become lighter in the process, or how much mass has the sun lost during its current lifespan of 4.5 billion years? We first convert the unit electron volt to joules:

$$1\,eV = 1,6 \times 10^{-19}\,J \tag{6.7}$$

According to Einstein

$$E = mc^2 \tag{6.8}$$

$$m = \frac{E}{c^2} \tag{6.9}$$

$$= \frac{1,6 \times 10^{-19}}{(3 \times 10^8)^2} = 1,8 \times 10^{-36}\,kg \tag{6.10}$$

Therefore, the sun loses per second through the neutrino flux[1] $1,8 \times 10^{-36} \times 1,8 \times 10^{38} \sim 100\,kg$. A year has about 3×10^7 s, so the sun has lost the amount of $4,6 \times 10^9 \times 3 \times 10^7 \times = 1,9 \times 10^{19}\,kg$ in the past 4.6 billion years. This corresponds to about 7×10^{-9} of its total mass, or 0.0000000007% of its total mass, so it is absolutely irrelevant.

Let's compare this value with the energy released during nuclear fusion: here, 26 MeV are released per reaction. Assuming there are as many nuclear reactions as released neutrinos (which is certainly true in order of magnitude), then the sun's mass loss since its formation through nuclear fusion is about 0.018% of its total mass, still extremely little.

There was a big surprise when measuring the neutrino flux coming from the sun. Only about 1/3 of the reactions that were expected were detected. The measured neutrino flux is therefore too small. This could have many causes:

---

[1] For simplicity, we assume that the neutrino mass is 1 eV.

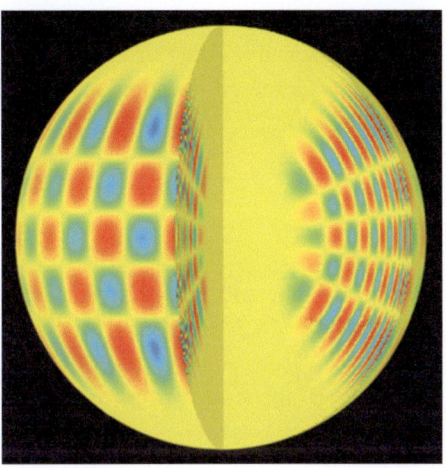

Fig. 6.11 Vibrations of the sun, the different colors indicate matter moving upwards (blue) or downwards (red)

- the pre-calculations of the neutrino flux are wrong,
- the neutrinos have a different cross-section,
- our ideas about the structure of the sun are wrong,
- something is wrong with the physics of neutrinos.

Elementary particle physicists blamed the astrophysicists: the ideas about the structure of the sun are not correct. Therefore, nuclear reactions occur less frequently than assumed, fewer neutrinos are produced. There were interesting, sometimes bizarre ideas. The interior of the sun could rotate faster than the rest; therefore, centrifugal forces occur, a lower temperature is needed to maintain balance against gravity. The gas pressure, which depends on the temperature in addition to the number of gas particles[2], can therefore be lower. A lower temperature inside the sun automatically leads to a lower nuclear fusion rate and therefore fewer neutrinos are produced. An exotic possibility would be to assume a black hole at the center of the sun, or that there could be exotic particles there—relics from the early days of the universe.

Fortunately, we have another way to look inside the sun: *helioseismology*. Our sun vibrates. Let's imagine a bell. Then we can infer the structure and composition of the bell from its sound. The same is true for the sun's vibrations. We even know more about the interior of the sun than about the interior of the earth. In Fig. 6.11 you can see a vibration pattern of the sun. This figure shows that the sun's vibration patterns spread from the surface into its interior. Red means matter moving inward, blue means matter moving outward. The basic period of the vibrations on the sun is about five minutes. The solar oscillations were discovered in 1960 by *R.B. Leighton*.

---

[2] Temperature is a measure of the average kinetic energy of the particles.

The propagation speed of a disturbance inside the sun can be estimated with the formula for the propagation speed of a sound wave in a gas. This is given by the relationship:

$$c_s = \sqrt{\frac{p}{\rho}} = \sqrt{\kappa \frac{RT}{M}} \qquad (6.11)$$

Here, κ is the adiabatic exponent, p is the pressure, T is the temperature, ρ is the density, $R = 8{,}31\,J\,mol^{-1}\,K^{-1}$ is the gas constant, M is the molar mass (mass/amount of substance). From this, we can see how the other quantities (pressure, density, temperature) result from the propagation speed.

▶ The interior of the sun can be explored by analyzing the sun's vibrations (helioseismology).

The propagation of different waves into the interior of the Sun is shown in Fig. 6.12. The larger the oscillation pattern, the deeper it penetrates into the interior of the Sun. From the surface to a depth of about 200,000 km, the Sun is convective, hot gas flows upwards, cools down at the surface and sinks back down. This process is similar to boiling water,

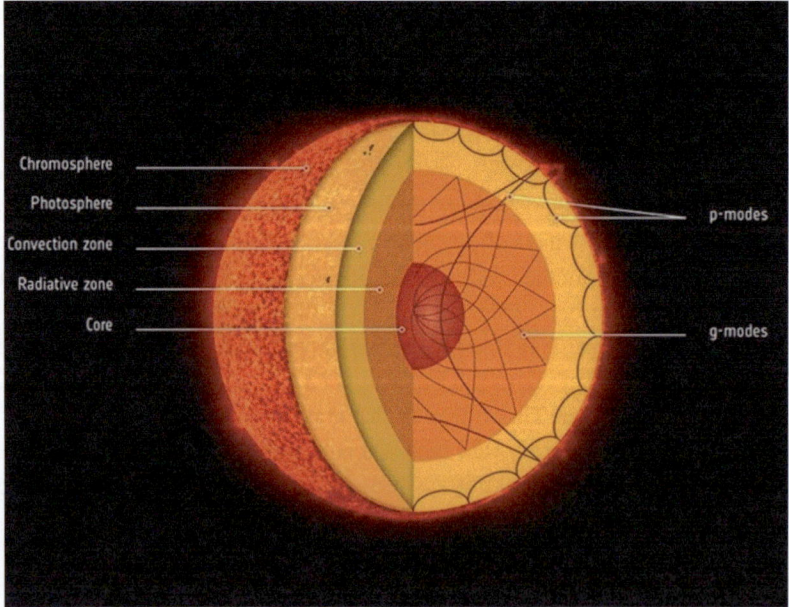

Fig. 6.12 Structure of the Sun and propagation of oscillations into the interior of the Sun

6.3 Where do Neutrinos from the Universe Come From

where gas bubbles form at the bottom of the heated pot rise upwards, etc. Below the *convection zone* lies the *radiation zone,* here energy is transported by radiation. The energy itself is produced in the *core area* by fusion. This produces high-energy gamma quanta, which are scattered by free electrons in the radiation zone after only about 1 cm. As a result, these quanta lose some energy with each scattering process and, before they reach the surface, become longer-wave quanta, i.e., for example, visible light that we observe. But there is a curiosity here: when we look at the Sun today, we see light quanta that were produced inside the Sun about 100,000 years ago, because that's how long it takes for these quanta to reach the surface.

▶ Helioseismology teaches us that the concepts of the structure of the Sun fit very well.

So if our models of the Sun agree with the results of helioseismology, something must be wrong with the physics of neutrinos. The solution is: *Neutrino oscillations.* Neutrinos actually come in three states: electron neutrinos v_e, muon neutrinos v_μ and tau neutrinos v_τ. In the fusion of hydrogen to helium, only electron neutrinos are produced. However, these oscillate and an electron neutrino can become a muon neutrino or a tau neutrino. However, only electron neutrinos can be detected with the detectors. Thus, solar physics has made an important contribution to the exploration of the properties of these mysterious particles. But there is another effect of these neutrino oscillations. These can only exist if the neutrino has a (albeit extremely small) rest mass[3]. Since there are many neutrinos in the universe, they could therefore make a certain contribution to the total mass of the universe.

▶ The problem of solar neutrinos is solved by neutrino oscillations.

6.3.2 Neutrinos from a Supernova Explosion

At the end of their lives, massive stars explode into a *supernova* and what remains is either a *neutron star,* or for even larger masses, a *black hole.* In such a collapse, about 10^{58} neutrinos are released. These possess high energies of 10–20 MeV. The neutrinos from the sun only have about 1/50 of these energies. The neutrinos are generated in the core area and therefore their observation gives us insight into the processes in the core of an exploding supernova. Currently, only neutrinos from the supernova explosion SN 1987A have been detected.

---

[3] For a long time, it was thought that neutrinos, like photons, have a rest mass of zero and can propagate at the speed of light.

Fig. 6.13 The supernova SN 1987A near the Tarantula Nebula in the Large Magellanic Cloud. ESO

The brightening of the supernova in the *Large Magellanic Cloud* near the so-called *Tarantula Nebula* can be seen in Fig. 6.13. The Large Magellanic Cloud is a companion galaxy of our Milky Way and is about 168,000 light-years away from us.

In Fig. 6.14, you can see the measured increased flux of neutrinos during the explosion of SN 1987A. With the three stations Kamiokande II, IMB and Baksan, a total of 25 neutrino events were measured. Only electron neutrinos were found, because the energy was below the production threshold for muon or tau neutrinos. These electron neutrinos are produced by the following process:

$$e^- + p \rightarrow \nu_e + n \qquad (6.12)$$

In Fig. 6.15, the development of the matter ring ejected during the explosion is shown. Initially, the ring becomes weaker, but then it is heated by the arrival of the expanding shock wave.

However, it was remarkable that initially no neutron star was found as a remnant of the explosion. The formation of a black hole could be ruled out, as the mass of the progenitor star was known. Only radio astronomy could then help. Indications of a neutron star that formed after the explosion can be found in Fig. 6.16. These observations were made with the ALMA radio telescope. The neutron star is not yet directly visible, as it is obscured by a dense cloud of dust that it heats up. This heating provides the detected radio emission of the object.

6.3 Where do Neutrinos from the Universe Come From

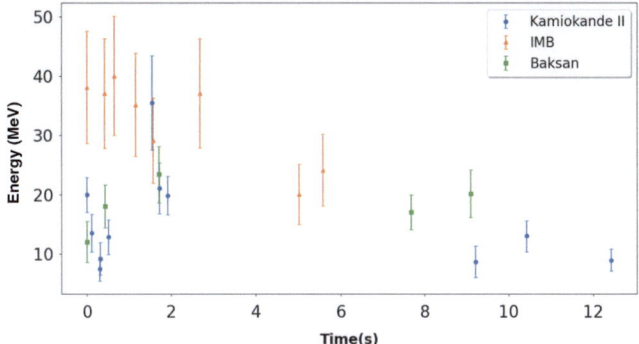

Fig. 6.14 Measurements of neutrinos during the explosion of SN 1987A

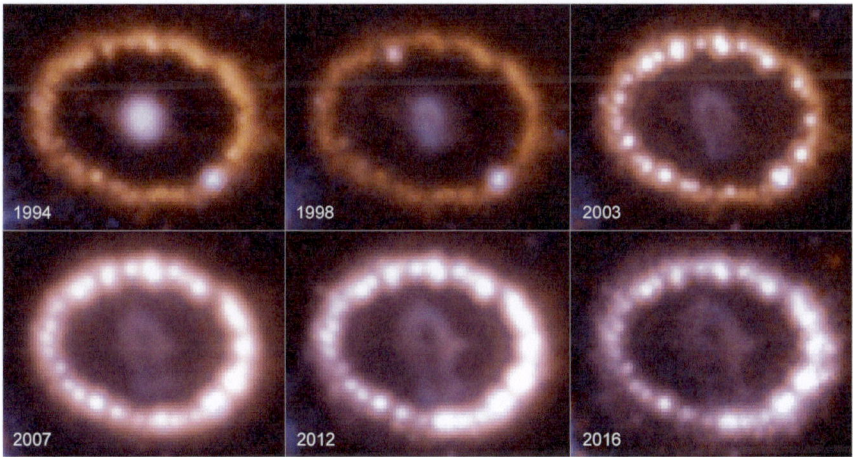

Fig. 6.15 The development of a ring around the supernova SN 1987A. NASA,ESA, R. Kirsher, B. Moore, P. Challis

Before the visible light from SN 1987A reached Earth, an increased neutrino flux was measured 2–3 hours earlier. The visible light is only transmitted after the shock wave, which was formed by the core collapse, reaches the surface of the star. Numerical models of a supernova explosion show that 99% of the energy of the collapse is radiated in the form of neutrinos. The total number of neutrinos is given as 10^{58}.

So neutrino astronomy can essentially provide a warning for a supernova explosion. With even more sensitive detection possibilities, it is hoped to observe such events also, for example, in the Andromeda galaxy.

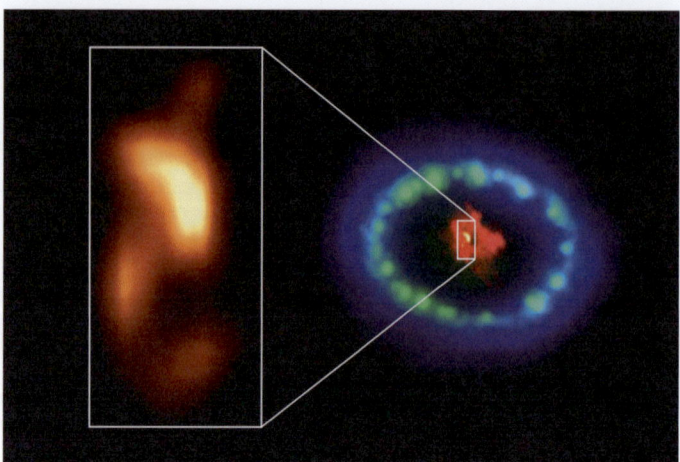

Fig. 6.16 Indications of a neutron star at the center of the exploded supernova SN 1987A. Clear matter jets can be seen, as would be expected from a neutron star. Credit: ALMA (ESO/NAOJ/NRAO), P. Cigan and R. Indebetouw; NRAO/AUI/NSF, B. Saxton; NASA/ESA

The Hubble Space Telescope

7

A telescope in space has long been the dream of astrophysicists. The advantages are undisturbed, weather-independent observation, there are no restrictions regarding wavelengths; the advantages are of course offset by a great technical effort as well as significantly higher costs and risks. As an illustrated book on this, for example, [11] is recommended.

7.1 Planning, Launch, Problems

7.1.1 First Ideas

We have seen that there are numerous large telescopes at climatically favored locations on Earth. These have the advantage of easy accessibility. However, the disadvantage is the Earth's atmospherere:

- The Earth's atmosphere causes turbulence.
- These turbulences affect the image quality, very complex technology is needed to get close to the theoretical resolution; e.g. adaptive optics.
- The Earth's atmosphere is only permeable in the optical range, in some narrow infrared windows, and in large parts of the radio range.

For this reason, astrophysicists wanted to position telescopes in space. The first space telescopes were primarily designed to observe objects in wavelength ranges that are not accessible from the Earth's surface.

Thus, the planning of the first space telescope named after E.P. Hubble (1889–1953) was created, which should have a larger mirror diameter and should also deliver images

Fig. 7.1 The Space Shuttle spacecraft. NASA

in the optical wavelength range. As early as 1946, the astrophysicist L. Spitzersuggested positioning a telescope in space. He wrote a publication titled "Astronomical Advantages of an Extra-Terrestrial Observatory" in which he presented the previously mentioned points as great advantages of a space-bound telescope. He concluded that a space telescope would far outperform an earthbound telescope in smaller dimensions. However, at that time nothing was known about the possibilities of correcting atmospheric disturbances using adaptive optics. In 1966, further studies on the subject of a large space telescope were commissioned by the US National Academy of Sciences. It was clear that such a project could only be realized in cooperation with the US space agency NASA. NASA also had its own studies on telescopes in space, but these were usually smaller.

In the mid-1960s, a very important decision was made for the construction of a large space telescope: they wanted to build a space shuttle, the *Space Shuttle* (Fig. 7.1); with this reusable shuttle, on the one hand, the observatory could be brought into space and, on the other hand, technical improvements and innovations to the instruments of such a space telescope could be carried out much more cost-effectively than with rockets that are only used once. The Space Shuttle was 37 m in size and had a wingspan of 24 m. Up to 8 astronauts could be brought into Earth orbit with it. The large external tank held 700 t of fuel, the two solid rockets less. It could bring material up to about 600 km above the Earth's surface.

But back to the Hubble Telescope. Time passed. In 1971, the *Large Space Telescope Steering Group* was then established by NASA. A committee that dealt with questions about such a project. So it was about a kind of feasibility study. Of course also about the money, what costs would such an instrument cause.

Fig. 7.2 The OAO-1 satellite was launched in 1966. NASA

7.1.2 Precursor Missions

Another milestone on the way to the large space telescope were the successful *Orbiting Astronomical Observatory Satellites, OAO* (Fig. 7.2). Most of these satellites operated in the UV range, with mirror diameters between 30.5 and 97 cm. The satellites were launched between 1966 and 1972. The *OAO-1* shown in the picture was not very successful, shortly after the launch in April 1966 there was a power failure and after three days the mission was officially aborted. The satellite was supposed to measure gamma radiation from cosmic sources. *OAO-2* was successful, it had 4 telescopes (around 30 cm) observing in different UV ranges. Among other things, it was discovered that comets are surrounded by a huge hydrogen cloud. The satellite was also known as "Stargazer". *OAO-B* was equipped with a 97 cm mirror, but there were problems shortly after the launch and the satellite burned up in the Earth's atmosphere. *OAO-3* was successful: the mission was also known as Copernicus. The satellite provided data from 1971 to 1981, and it was equipped with an 80 cm mirror telescope for UV observations (Fig. 7.3).

In 1983, the *IRAS* (Infrared Astronomical Satellite) was launched, which observed in the infrared, with a mirror diameter of 60 cm. In Fig. 7.4 you can see a false color image of the Milky Way obtained from IRAS data. The different colors correspond to: blue = $12\,\mu m$, green = $60\,\mu m$ and red = $100\,\mu m$. Blue and white objects are warmer, red ones are colder. You can see the band of the Milky Waye.

The next step was financing. Originally, a 3-m mirror was considered, which would have cost around 400 million US dollars at the time. Due to the high costs, the proposal was rejected by the House Appropriations Committee. A small group of well-known astrophysicists then formed and lobbied intensively. In addition, partners were sought

Fig. 7.3 The OAO-1 satellite was launched in 1966. NASA

Fig. 7.4 The IRAS satellite mapped the entire sky in the infrared range, the band of the Milky Way can be seen. NASA

and ESRO (a precursor organization to today's European Space Agency ESA) was won over; in return, the later ESA received observation time on the telescope. To further reduce costs, a mirror with a diameter of 2.4 m was chosen, which halved the costs, and in 1979 the budget funds were approved by the US Congress. By the way, two identical mirrors were manufactured in case one should fail. Both passed the final quality testing. The launch of the telescope was initially scheduled for 1983, but the date could not

7.2 The Great Disappointment

Fig. 7.5 The deployment of the Hubble Telescope with a crane at an altitude of 600 km above the Earth's surface. NASA

be met due to various problems. The next targeted launch date was October 1986. But unfortunately, the Challenger disaster occurred in January 1986. The space shuttle *Challenger* was to be launched into Earth orbit with a solid rocket booster, but it exploded, killing all seven astronauts. The investigation into the circumstances leading to this greatest tragedy in manned space flight to date took a very long time, and therefore the launch was delayed again. On April 24, 1990, however, the time had finally come. The Space Shuttle mission designated STS 31 launched on schedule at 12.31 with the space shuttle *Discovery*, which had the Hubble Telescope stowed as payload. The telescope was deployed at an altitude of 611 km above the Earth's surface. A crane developed by Canada was used for this purpose (Fig. 7.5).

7.2 The Great Disappointment

7.2.1 An Error in Optics…

Space missions take a very long time, as we have seen. The first ideas for a space telescope existed several decades before the project could be realized. After its release into Earth's orbit at 600 km on April 24, 1990, everyone was naturally very excited about the first images. And then there was a great disappointment. Although the main mirror underwent a thorough quality control, the images were blurry. In Fig. 7.6 you can see the image of a star, it is the first image taken with the Hubble Telescope (HST). The star in

Fig. 7.6 The first image from the Hubble Telescope; instead of sharp stars, only a heavily blurred image is recognizable. NASA

the middle should be depicted as a point without scattered light. The telescope therefore shows the error of *spherical aberration*.

Of course, attempts were made to reconstruct how this error occurred. It was probably a communication problem or the quality assurance was not carried out by an independent team. In addition, there were no clearly defined criteria according to which the final tests could be rated as passed or failed. The mirror built into the telescope was made with a new manufacturing technique, the replacement mirror left on Earth with conventional techniques. It was considered to replace the mirror in the HST with the reserve mirror, but this would be technically extremely complex, so better methods were sought to correct the faulty optics.

7.2.2 The Error is Corrected

Through many tests and computer simulations, the *COSTAR* system was developed to correct the main mirror error before the light hits the various cameras and instruments. COSTAR is the abbreviation for Corrective Optics Space Telescope Axial Replacement. It took two and a half years before this system could be installed in the Hubble Telescope during a service mission. In Fig. 7.7 you can see the assembly. Astronaut Kathryn C. Thornton lifts the Corrective Optics Space Telescope Axial Replacement (COSTAR) before installation on the Hubble Space Telescope (HST). Thornton is anchored to a foot restraint at the end of the RMS arm (Remote Manipulator System). Astronaut Thomas D. Akers, who assisted in the COSTAR installation, is standing in the lower left.

The relief was great when the first sharp images were received after the commissioning of COSTAR. An example is shown in Fig. 7.8.

7.2 The Great Disappointment 153

Fig. 7.7 Repair in space: Only by installing a corrective optic could sharp images be obtained with the HST. NASA

Fig. 7.8 Comparison of an image from the Hubble Telescope before commissioning (left) and after commissioning of the COSTAR correction system (right). NASA

7.3 Instruments of the Hubble Telescope

7.3.1 What Does a Space Telescope Look Like?

In itself, one would somewhat uncharmingly describe the Hubble Space Telescope as a cylindrical tube that has a length of 13.2 m and whose diameter reaches up to 4.3 m. The whole thing weighs 11.1 t. Also striking is a flap at the front end of the telescope through which the light comes in. This flap has a diameter of 3 m (Fig. 7.10). On the outside, it is coated with a layer that reflects sunlight. The system as a whole has very light-sensitive sensors, and as soon as the sun falls at an angle of less than 20° from the alignment axis of the telescope, the flap automatically closes to protect the devices. The telescope is powered by solar cells. The silicon-based modules deliver a power of about 4500 W and the dimensions of the cells are each 12.1 m × 2.5 m, the weight was 7.7 kg each. The first modules had to be replaced during the service mission SM1, however, as they deformed due to the strong temperature changes. When the telescope is exposed to sunlight again during its orbit around the Earth, the temperature changes within a very short time from −100 degrees to +100 degrees. During the time when the telescope is in the Earth's shadow, the sun's rays do not reach the solar cells and the energy stored in nickel-hydrogen accumulators is used for supply. These accumulators are designed so that the telescope can be operated for about 7.5 h without sunlight (Figs. 7.9 and 7.10).

The data cannot be sent to Earth in real time, so intermediate storage is necessary. Many readers may not even remember that there used to be magnetic tapes for recording data. In the early years of the Hubble Telescope's operation, two tape drives with

Fig. 7.9 The Hubble Telescope with the flap open. NASA

7.3 Instruments of the Hubble Telescope

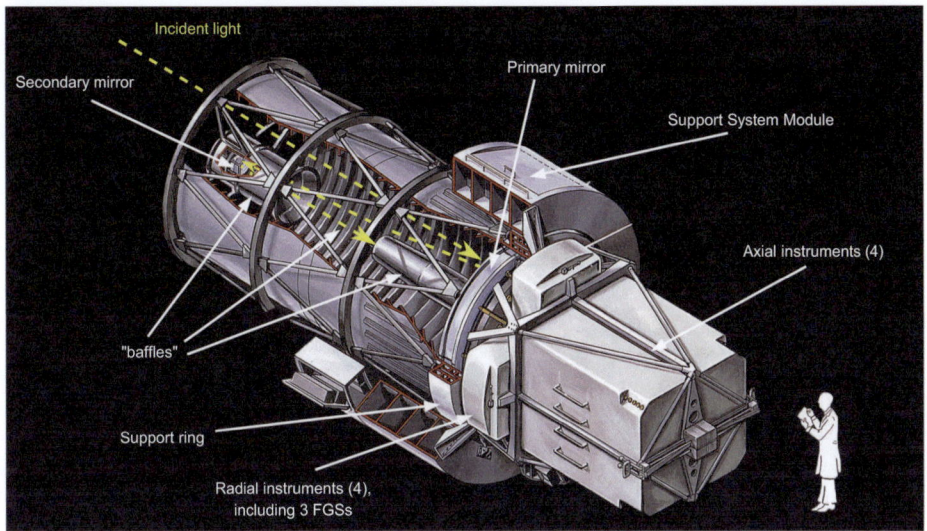

Fig. 7.10 The Hubble Telescope: Design. NASA

1.2 GBit data storage capacity each were available, each of the drives weighed 9 kg. Consider the amount of data you can store on a simple USB stick today—a multiple of that at a weight of a few grams and extremely small dimensions. During the service mission SM2, these drives were replaced by two SSD storage devices (each 12 GBit).

For *communication* there are two highly sensitive antennas (HGAs, high gain antennas). The HGAs have 2 parabolic mirrors and have a diameter of 1.3 m. They can be aligned as desired to ensure optimal data forwarding. The data rate is about 1 MBit/s; therefore, 120 GBit of data can be transmitted per week.

To precisely align the telescope to any point in the sky, it has a *Pointing Control System, PCS*. With it, the telescope can be aligned to an accuracy of 0.01 arcseconds in the sky and track an object for at least 24 h with an accuracy of 0.007 arcseconds. Tracking is necessary because the telescope is moving around the Earth. This accuracy corresponds to the angle under which one would see a 10-cent coin at a distance of 6000 km! *Gyroscopes* are used for alignment and stabilization.

The 2.4 m large primary mirror (Fig. 7.11) was cooled from a temperature of over 1000 degrees to room temperature for three months to avoid any tensions that might occur during too rapid cooling. Only then could the grinding begin. The reflective surface is only 3.8 cm thick, behind which is a honeycomb structure, about 25 cm thick, that supports this thin mirror. In this way, weight could be saved, the mirror weighs only 818 kg. The secondary mirror has a diameter of 30 cm. The focal length of the system is a maximum of 57.6 m. The light is guided by the secondary mirror through a 60 cm hole in the main mirror to the instruments located behind it.

Fig. 7.11 The primary mirror is being ground. NASA

7.3.2 Instruments on Board the HST

Let's start with the *High Speed Photometer, HSP*. This device can measure the brightness and polarization of variable stars up to 10000 times per second on several channels (e.g. the Cepheids).

When it turned out that the main mirror had been ground incorrectly, and the correction system COSTAR had to be accommodated, the HSP fell victim to space constraints and was replaced by COSTAR.

With the *Faint Object Camera, FOC,* two measurement channels were used and the image of the main mirror was enlarged 2 or 4 times, so we had an f/48 or f/96 system. Due to the high focal length, only a small sky field could be observed, namely:

- $44 \times 44''$ or
- $22 \times 22''$

This would be roughly as if you could only see Jupiter or Saturn in the image field, which then fills the entire field. The detectors were sensitive in the range of 1110 nm to 650 nm. The image size was 512 by 512 pixels and the resolution was 0.014 per pixel.

With the *Faint Object Spectrograph (FOS)* one could break down the light of the objects and determine the composition of the objects from the analysis of the spectrum or determine the speeds of gas clouds around black holes from the Doppler shifts. This device was replaced by NICMOS during Service Mission 2.

7.3 Instruments of the Hubble Telescope

Fig. 7.12 The Advanced Camera for Surveys, ACS. NASA

NICMOS stands for *Near Infrared Camera and Multi-Object Spectrometer*. With it, observations were made in the wavelength range 800–2500 nm. Since this wavelength range is in the near infrared, cooling was necessary. Liquid nitrogen was used for cooling, initially there was a supply of 109 kg, which lasted for 2 years. Then a closed cooling system was installed during a service mission. At the end of 2008, the system no longer worked after a software update (basic rule in physics: never do a software update!!). But until then, for example, objects could be studied that remained hidden in visible light behind dust clouds.

The *Advanced Camera for Surveys, ACS* (Fig. 7.12), is a camera for observing large sky areas. In addition to the visible range, it can also observe in UV and near infrared. This device was installed during Service Mission SM 3 B and replaced the previously used Faint Object Camera, FOC. Fig. 7.13 shows astronauts during the dismantling of the FOC and installation of the ACS.

For the examination of the objects, there were three channels:

- HRC, High resolution channel—with this, the objects could be studied in high resolution, i.e., in detail,
- WFC, wide field channel, with this, wide-angle shots could be taken and a
- SBC channel (solar blind channel) for observations in UV.

The HRC offered another possibility: weak objects near bright objects could be observed. This is usually hardly possible due to the high contrast difference. However,

Fig. 7.13 Astronauts dismantle the FOC and replace it with ACS. NASA

Fig. 7.14 The beam path in the WFC of the Advanced Camera for Surveys. StScI handbook, NASA

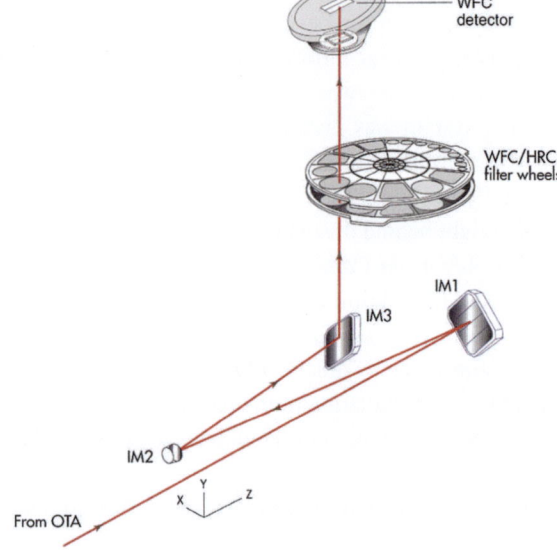

a mask was used here, with the help of which the bright object was dimmed. The HRC failed in 2006 due to a defect and could not be repaired. In Fig. 7.14, the beam path in the WFCC is outlined. A filter wheel can be seen, with which several filters can be positioned in the beam path. The light comes from the OTA (optical tube assembly) into the system. The mirrors are coated with silver to ensure high reflection in the optical and UV rangeensure.

7.4 Some Observation Results With the Hubble Telescope

7.4.1 Solar System

Modern insights about the solar system can be found in books [2] as well as [8] and [10].

In Fig. 7.15, one can see an image taken with the ACS in HRC mode (Advanced Camera System in High Resolution Channel mode) of the planet *Mars*. Never before could so many details be recognized on our outer neighboring planet from Earth. The smallest recognizable structures are about 8 km in size. The black arm with the disk is the *coronagraph*, which could cover this star for observing weak objects near a bright star.

Another example of a high-detail observation of a planet is shown in Fig. 7.16: the planet *Jupiter*. Jupiter was at a distance of 668 million km from Earth at the time of the recording. In addition to the Great Red Spot (bottom left), one can see a smaller red spot (bottom right, also referred to as the junior red spot). The image was taken with the Wide Field Planetary Camera and shows details up to about 130 km. In the bright zones and dark bands parallel to the equator of Jupiter, which rotates around its own axis in less than 10 hours, there are strong winds with speeds of up to 400 km/s.

In Fig. 7.17, one can see Saturn with two moons casting a dark shadow on its surface as they pass (transit) the planet . The largest Saturn moon, *Titan* (top right), is easiest to recognize. The moon itself glows red and its shadow on Saturn is clearly visible as

Fig. 7.15 An image of the planet Mars with the HST, ACS-HRC. The image was taken on August 24, 2003. NASA

Fig. 7.16 Image of the planet Jupiter. The colors are created by different molecules in the atmosphere of the giant planet. The image was taken on April 3, 2017. NASA/ESA, Simon. A

a large black spot. Titan is larger than the planet Mercury and has a very dense atmosphere, and there are lakes of liquid hydrocarbon compounds on its surface. Upon closer inspection, one can see a small black spot about 1/3 from the left edge of Saturn, slightly north of Saturn's rings, and immediately to the left of it a bright spot. The bright spot is the moon Mimas, which casts a much smaller dark shadow on Saturn's surface. To the left of Saturn, one can see the bright moon *Dione*.

7.4.2 Stars, Nebulae

One of the most famous images from the Hubble Telescope are the so-called *Pillars of Creation,* a dust formation in the Eagle Nebula, referred to as M16 in the Messier Catalog. In Fig. 7.18, one can see a wide-field image of the Eagle Nebula, and in Fig. 7.19, a detailed image showing the so-called Pillars of Creation.

The *Eagle Nebula* is about 7000 light years away from us and is illuminated by young hot stars (so-called emission nebula). The two dust columns visible in the detailed image are about 10 light years long. The stars are less than a million years old, so very young, some stars may only be 50,000 years old. The image with the HST was taken in

7.4 Some Observation Results With the Hubble Telescope

Fig. 7.17 Saturn with moons casting shadows on its surface. HST-WFPC, February 24, 2009. NASA/ESA, M.H. Wong, C. Go, STScI

Fig. 7.18 The Eagle Nebula, M16. ESO

1995. The shape of the columns could have been created by a supernova explosion that occurred about 2000 years ago. Due to the relatively short distance of 7000 light years, this supernova should have been noticeably bright in the sky even in daylight. Since the light is on its way to us for 7000 years, it could even be that this formation no longer exists now. In Fig. 7.20 an image of the Eagle Nebula taken with the Spitzer Space Telescope is shown. The known structures visible in visible light are colored green, particularly hot areas are red. The position of the Pillars of Creation or a reduced image with the HST is also marked. The Spitzer Space Telescope, SST (originally referred to as SIRTF, Spitzer Infrared Telescope Facility), was fully active from 2003 to 2009, it was cooled down to 3 K by liquid helium; thereafter the temperature was around 31 K, and the telescope operated until 2020.

7.4 Some Observation Results With the Hubble Telescope

Fig. 7.19 The Pillars of Creation in the Eagle Nebula, M16. NASA/HST

Fig. 7.20 The Eagle Nebula, captured in infrared with the Spitzer Space Telescope. NASA/SST

In the outer regions of our Milky Way, in the so-called *Halo* are the *Globular clusters,* which contain several 100,000 stars. They are among the oldest objects in the universe, so they contain very old stars. These stars are also referred to as stars of the *Population II*. They contain almost no elements heavier than helium. Our sun belongs to Population I, so it consists of material that has already been enriched with metals by the explosion of earlier stars. Fig. 7.21 shows the globular cluster NGC 6355, which is about 50,000

7.4 Some Observation Results With the Hubble Telescope

Fig. 7.21 The globular cluster NGC 6355. ESA/Hubble & NASA, E. Noyola, R. Cohe

light years away from us[1]. As a result of the gravitational forces between the densely packed stars, globular clusters stay together. Due to the high star density, a planetary system in a globular cluster would not remain stable for long. Estimates show that a planet at a distance of 150 million km from its parent star (equivalent to the Earth-Sun distance) would be catapulted out of the system within about 100 million years due to disturbances from nearby stars.

To better imagine the density of stars in a globular cluster, let's consider the following model. If the diameter of the sun is about the size of a cherry, then the nearest stars would be about 300 km away from us. In a globular cluster, where the star density is about 10,000 times greater, the distance between the stars is still $300000/10000 = 30$ m. Stars, as big as cherries, would still be 30 m apart in the densest parts of a globular cluster. The probability that two cherries 30 m apart would collide is still low.

In the outer regions of our Milky Way, there are about 150 globular clusters. There are also galaxies with more than a few hundred globular clusters (e.g., the galaxy M87 contains several 100,000 such clusters).

In Fig. 7.22 one can see the *Crab Nebula*, M1. A star appeared as a bright object in the year 1054, which was also visible in the daytime sky. Today, we find the Crab Nebula at exactly this location. It is the remnant of a supernova explosion. At the center of the nebula, which has now grown to about 10 light-years, there is a neutron star only about 10 km in size, the nebula itself consists of the shell ejected during the explosion.

---

[1] NGC stands for New General Catalogue, a directory of more than 7000 celestial objects.

Fig. 7.22 The Crab Nebula M1. NASA, ESA, Hubble, J. Hester, A. Loll (ASU)

7.4.3 Galaxies

In Fig. 7.23, we see an image of the galaxy NGC 3370. This image, obtained with ACS in HRC mode, shows so many details that individual Cepheids can be measured in this galaxy. We know what this means: this allows the distance of this galaxy to be precisely determined. A few years before the image was taken, a supernova exploded in this galaxy. Since all supernovae shine with approximately the same brightness, their easily measurable apparent brightness gives the distance. Both methods of distance determination yielded roughly the same result: the galaxy is 98 million light-years away from us.

In Fig. 7.25, we see the *galaxy cluster* Abell 370. This contains hundreds of galaxies bound by gravity. The cluster is about 4 billion light-years away from us. Objects behind this cluster are distorted by the gravitational lens effect . The large mass of the cluster curves space-time (see chapter on General Relativity) and therefore these objects appear distorted. In the image, we also see thin lines, some of them S-shaped. These are traces

7.4 Some Observation Results With the Hubble Telescope

Fig. 7.23 An image of the galaxy NGC 3370 with the HST, ACS-HRC. The image section is about 100,000 light-years in diameter at the distance of the galaxy. NASA

of asteroids in our solar system, only 200 million km away from us. These strange traces are created by overlaying many images of the galaxy cluster with long exposure times. The arcs are nothing more than parallaxes. The Hubble Telescope moves around the Earth, so it sees the relatively close asteroids from different angles. Another effect that leads to a parallax results from the movement of the Earth around the Sun (Fig. 7.24).

The angular size of the image in the sky is only 2 arc minutes.

Let's stay with this galaxy cluster. Here, for the first time, a supernova was observed that was imaged three times due to the gravitational lens effect. The image Fig. 7.26 on the left shows the galaxy cluster and on the right, we then see an enlarged section, a total of 4 images. In the first image, we see a combined image from 2011–2016, then the image from 2010, when the supernova was discovered. The difference images in black and white or color are then given. Note that the three objects labeled "1,2,3" are different images of the same object, the supernova!

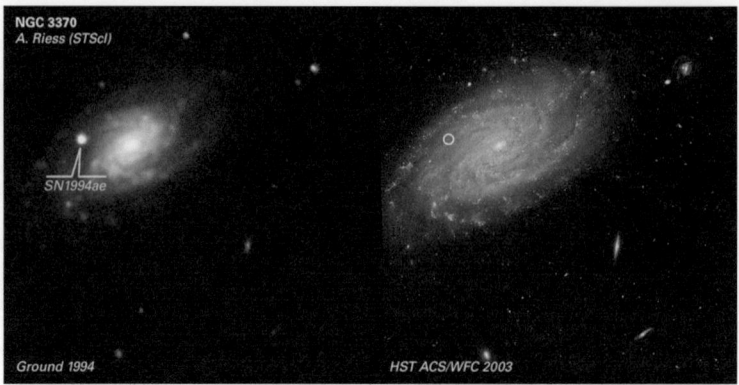

Fig. 7.24 Two images of the galaxy NGC 3370: on the left with a telescope on Earth with a supernova, on the right with the HST, ACS-HRC. M. Riess, NASA

7.4.4 A Look into the Early Universe: The Hubble Deep Field

The Hubble Deep Field project, HDF, refers to an extremely long-exposure image of a region of the sky. Deep Field here means that one wants to look as deep into space as possible, i.e., observe extremely distant galaxies. As already mentioned, due to the finite speed of light, a look into the depths of the universe is always simultaneously a look into the past. So when we see distant galaxies, we are seeing galaxies in the early universe. Several Deep Fields have been captured.

Which areas of the sky are suitable for such Deep Field images? To observe distant galaxies undisturbed, it is advantageous to select the sky field in such a way that as few stars as possible that belong to the Milky Way are visible in the images. These relatively nearby stars shine brightly compared to the extremely distant galaxies. A central question in cosmology is indeed when galaxies existed in the universe. Such Deep Field images should clarify whether there were already galaxies in the early universe.

The field captured between December 18 and 28, 1996, measures only 144 arcseconds in the sky, which is about four times the apparent diameter of Jupiter's planetary disk. Or to put it another way: imagine the size of a tennis ball at a distance of 100 meters, then you have an impression of the size of the sky field of the Hubble Deep Field. The final image consists of an addition of 342 individual images. Adding individual not too long-exposed images achieves the same effect as if you only take a very long-exposed image. Due to the movement of the telescope around the earth, the recording time of an individual image is of course limited. The image was taken with the Wide Field Planetary Camera 2.

The data from the HDF led to 400 scientific publications and showed that there were already galaxies in the early universe. In Fig. 7.27 you can see the HDF. Almost all objects in the image are galaxies, each with up to several hundred billion stars. Stars

7.4 Some Observation Results With the Hubble Telescope

Fig. 7.25 A galaxy cluster that clearly leads to a gravitational lens effect, which can be seen in the distorted images of the galaxies. The thin curved lines are traces of asteroids. HST, ACS-HRC. NASA/HST, J. Lotz STScI and the HFF team

of our Milky Way, which happen to be in the line of sight, can be easily recognized: they show a cross-shaped structure or spikes. This is caused by the diffraction of a point light source at the grid construction for holding the secondary mirror of the telescope. It is estimated that the galaxies visible in this image are up to 12 billion light years away from us.

In Fig. 7.28 you can see the southern Hubble Deep Field, here a region in the southern starry sky was selected (it bears the designation HDF-S). Also here the total exposure time was 10 days and the image was created in September/October 1998. An area in the constellation Tucan was selected which, like the northern Hubble Deep Field, is far enough away from the galactic plane. This is important because in addition to the stars,

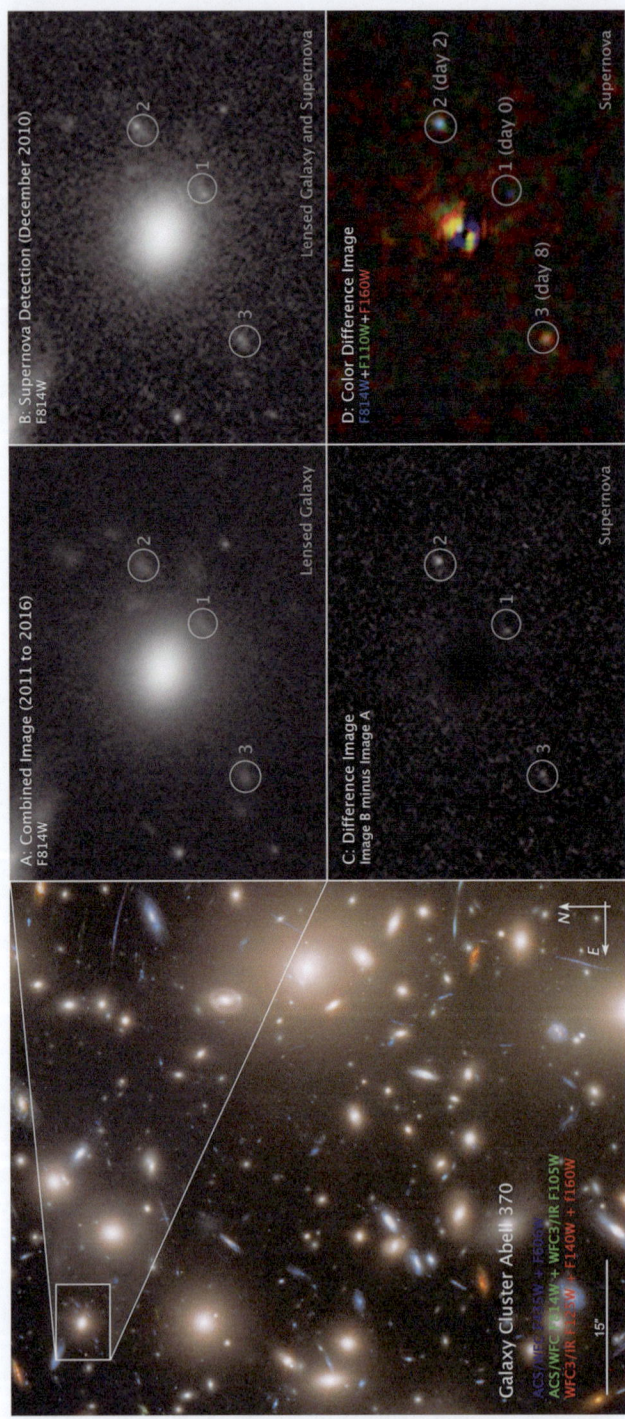

Fig. 7.26 A triple image of a supernova in the galaxy cluster Abell 370 caused by a gravitational lens. HST, ACS-HRC. NASA/HST, Wenlei Chen (UMN), Patrick Kelly (UMN), STScI

7.4 Some Observation Results With the Hubble Telescope

Fig. 7.27 The Hubble Deep Field provides a view into the early universe in visible light and shows the first galaxies. Robert Williams (NASA, ESA, STScI)

the interstellar matter (consisting of gas and dust) is also concentrated towards the galactic plane. The light of distant galaxies is greatly attenuated by the dust.

The universe is homogeneous and isotropic. So it doesn't matter in which direction we look. Proof of this assumption, which greatly simplifies the field equations that describe the relationship between matter and space curvature, is that the images of the northern and southern Deep Fields are similar.

In Figs. 7.29 and 7.30, one can see the UDF, the Hubble Ultra Deep Field. One can identify about 100 small point-like galaxies in this image, which appear reddish. The color is related to their high redshift. The Hubble Law states, that the further a galaxy is from us, the greater its redshift becomes. These highly redshifted galaxies are only about 800 million years old. The image comes from a region of the sky southwest of the constellation Orion in the constellation Fornax. The size of the sky field corresponds to that of a needle's eye through which one looks when held at a distance of one meter to the sky. It was created from individual exposures between September 3, 2003, and January 16, 2004. The total exposure time was about more than 11 days. The ACS camera was used as well as NICMOS, the Near Infrared Camera. In the image, one can see on the right in the zoomed-out small images, how point-like galaxies appear in the infrared. These are extremely redshifted.

Fig. 7.28 The Hubble Deep Field South was taken on the southern starry sky for the Earth's surface. NASA/the Hubble Space Telescope

The high redshift of the galaxies means that one must compare the image of these galaxies with corresponding images of nearby galaxies in ultraviolet light.

The *Extreme Deep Field,* EDF (Fig. 7.31) was created from a southern sky region and the images from a period of 50 days were superimposed. It was published on September 25, 2012. It shows galaxies that were formed only 450 million years after the Big Bang. The image shows more than 5000 galaxies.

7.4 Some Observation Results With the Hubble Telescope

Fig. 7.29 The Hubble Ultra Deep Field. NASA/the Hubble Space Telescope

Fig. 7.30 The sky section of the Hubble Ultra Deep Field. NASA/the Hubble Space Telescope

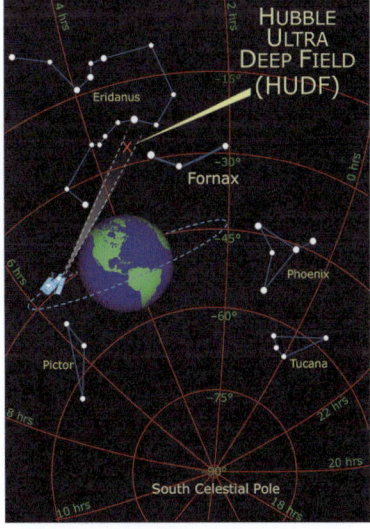

Fig. 7.31 The Hubble Extreme Deep Field. NASA; ESA; G. Illingworth, D. Magee, and P. Oesch, University of California, Santa Cruz; R. Bouwens, Leiden University; and the HUDF09 Team

The James Webb Telescope

8

The Hubble Telescope, which has been in space since 1990, has aged despite several service missions. Some of the instruments have been renewed, but at a certain point, the instrument as a whole is no longer state-of-the-art. Added to this is the rather small mirror diameter of 2.4 m. Therefore, a successor project was soon sought after the commissioning of the Hubble Telescope.

8.1 Planning

8.1.1 First Ideas for the Successor to the Hubble Telescope

The development of the James Webb Telescope, JWST (James Webb Space Telescope), started around 1996 as a joint project of three space organizations: the US NASA, the Canadian CSA, and the European ESA. In addition to the Hubble Telescope, it was also intended to replace the Spitzer Telescope, which operated in the infrared, and to combine and improve the advantages of both space telescopes:

- larger mirror diameter,
- Observation in the infrared.

Initially, the project was known as the Next Generation Telescope. The launch was planned for 2007. The construction costs were estimated at 3 billion US$ and were secured. There were delays, but the costs for five years of operation were secured in 2014, and the launch was expected for 2018. The transport into space was to be carried out with an *Ariane-5 rocket*. This rocket of the European Space Agency ESA has been in operation since 1996. It has a height of 50 m and a launch mass of 780 t. Of the more than 100 launches (as of January 2022), there were five failures. The rocket can carry more than 6 t of payload into a *geostationary orbit*. What is a geostationary orbit? At this distance of about 36,000 km from the Earth, the orbital period of a satellite equals the rotation period of the Earth. From the Earth's perspective, the satellite appears to stand still in the sky.

Figure 8.1 shows the Ariane-5 rocket with the JWST before the launch. If you, dear reader, would like to transport something into space with the Ariane-5: a launch costs about 200 million EUR, most launches are carried out on behalf of other organizations. The development of this rocket cost about 6 billion EUR.

The successor, the *Ariane-6* is supposed to be more cost-effective: a launch should then only cost 70 million EUR.

8.1.2 Difficulties

The main difficulty in the realization of the James Webb Telescope were the rapidly rising costs. The main mirror consists of 18 segments (Fig. 8.2), which were completed in February 2016 and the first optical tests in Earth laboratories took place half a year later. Initially, the telescope was to be transported into space in the spring of 2019, but the date was postponed to May 2020. There were problems with the control thrusters and the

Fig. 8.1 The Ariane-5 rocket launched the James Webb Telescope into space. ESA

8.1 Planning

Fig. 8.2 The main mirror of the James Webb Telescope, composed of 18 segments. NASA, JWST, ESA

deployment of the sunshield. The sunshield was necessary to protect the infrared-operated telescope from heating by solar radiation. In terms of costs, by 2019 it had reached 9.6 billion US$. It was clear that it would be the most expensive astronomical observatory to date.

The contribution of the European Space Agency ESA was around 300 million EUR. The following countries are involved: Belgium, Denmark, Germany, Finland, France, Greece, Great Britain, Ireland, Italy, Luxembourg, Netherlands, Norway, Austria, Portugal, Spain, Sweden, Switzerland, and the Czech Republic. The Canadian CSA was responsible for the Fine Guidance Sensor and the Near Infrared Slitless Spectrograph and also provided personnel for the operation of the Telescope.

8.1.3 The Launch

In September 2021, the telescope was tested again. All tests were positive and it was packed into a special container and loaded onto a ship at the end of September 2021 and sent to the Guiana Space Centre near Kourou. The complicated loading of this so far most expensive astronomical instrument is shown in Fig. 8.3. There was fear of pirates who might try to hijack the telescope to extort ransom. Therefore, nothing about the shipment was made public.

Fig. 8.3 The James Webb Telescope is loaded into a special container for transport by ship. NASA/Ch. Gunn

8.2 Structure and Instruments of the JWST

8.2.1 Construction of the Telescope

The James Webb Telescope, JWST, is similar in design to the Hubble Telescope, but its main mirror, due to its size of 6.5 m, is not made of a single mirror but of 18 segments. The light is reflected from the main mirror, M1, to a secondary mirror, M2, from where it hits another mirror M3 and then onto a fine adjustment mirror M4. The instruments are located in the image plane. The beam path is sketched in Fig. 8.4.

Figure 8.5 shows a size comparison of the JWST with the Hubble Telescope and the Spitzer Telescope. For the launch, the telescope, including solar cells and sun shield, had to be completely folded. It was a critical moment whether all components would unfold completely automatically in space (Fig. 8.6).

The launch mass was 6200 kg. The launch took place on December 25, 2021. On January 24, 2022, the telescope arrived safely at its destination, the Lagrange point L_2, at.

8.2 Structure and Instruments of the JWST

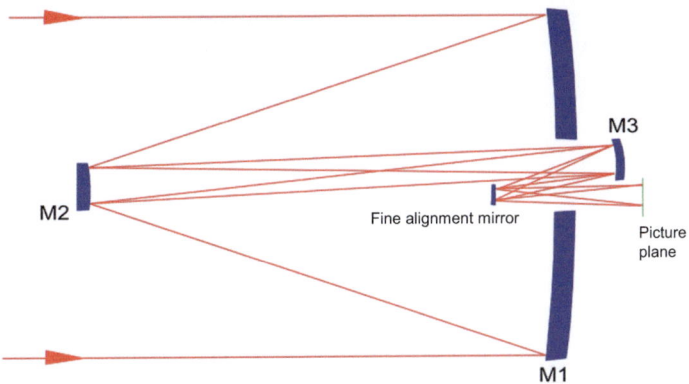

Fig. 8.4 The optical design of the James Webb Telescope NASA

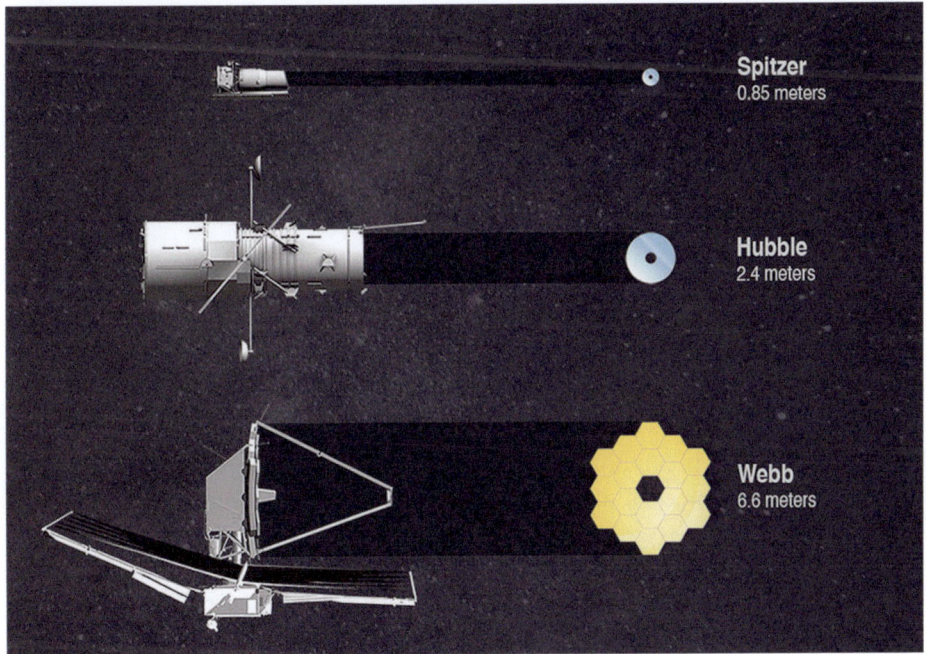

Fig. 8.5 Size comparison JWST with Hubble and Spitzer Telescope. NASA

8.2.2 Where is the James Webb Telescope located?

After the launch on December 25, 2021, the telescope began its journey to the Lagrange point L_2. Unlike the Hubble Telescope, which orbits the Earth once every about 90 minutes at a distance of only 600 km from the Earth's surface, this point is about 1.5 million

Fig. 8.6 The completely folded telescope. NASA

km away from the Earth. That is four times the Earth-Moon distance. This positioning of the telescope offers several advantages:

The so-called *Lagrange points* are equilibrium points in a so-called *restricted three-body problem.* In this problem, there are two masses, in our case the Earth and the Sun, as well as a third mass, which however does not influence the movement of the other two primary masses. If you bring a third mass, such as a satellite or a telescope, to one of these points, the forces of the Earth and Sun cancel each other out, and the object is stably parked.

The celestial mechanic *J.L. Lagrange* (1736–1813) pointed out the existence of these equilibrium points, there are five Lagrange points. In Fig. 8.7 their position is sketched. In the figure, the Sun is yellow and the Earth is blue. However, these considerations also apply, for example, to the Sun-Jupiter system and the group of *Trojan asteroids,* which are located near L_4, L_5 in this system.

However, only the points L_4, L_5 are truly stable, meaning that small disturbances return the object to its original position. The other points L_1, L_2, L_3 are unstable. Small disturbances on the satellite cause it to move further and further away from its position. Such disturbances are to be expected, the Earth and the Sun are not the only celestial bodies in the solar system. We have, for example, neglected the Moon as well as the influence of the neighboring planet Venus or Jupiter, the largest planet in the solar system.

8.3 Basic Supply and Instruments

Fig. 8.7 The Lagrange points $L_1, ..., L_5$. The stable points L_4, L_5 form an equilateral triangle with the Sun (yellow) and Earth (blue)

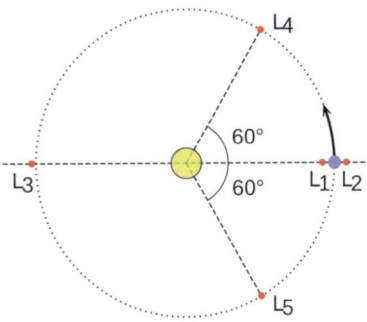

Fig. 8.8 The position of the JWST at the Lagrange point L_2

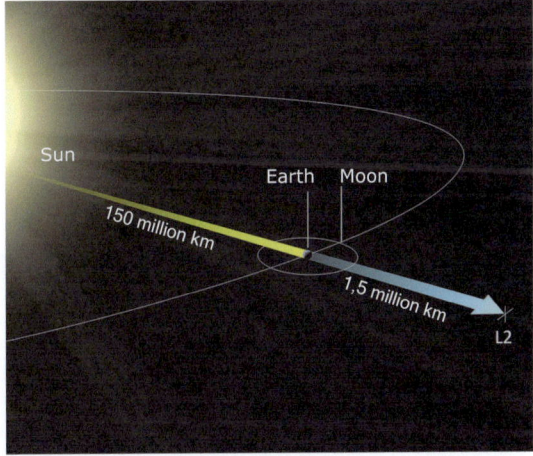

Figure 8.8 shows the position of the Lagrange point L_2 in the Sun-Earth system. The James Webb Telescope is in orbit around L_2.

▶ The James Webb Telescope is located at the Lagrange point L_2, a stable location from where it can observe the sky undisturbed.

8.3 Basic Supply and Instruments

8.3.1 Supply Unit, Power Supply

The technology for the basic functions of the telescope is housed in the supply unit, the English term is "Spacecraft Bus". In space, there is no atmosphere protecting against short-wave radiation and the probe is also outside the Earth's magnetic field, which protects us from charged particles. Therefore, the components of the supply unit are appropriately protected. In Fig. 8.9 you can see the supply unit (box-shaped), which is mounted below the sunshield.

The power supply is provided by solar panels that were deployed in space. These deliver a power of 2000 W. There are also accumulators. During planning, the aging of the solar cells (which progresses faster in space due to the intense short-wave solar radiation) and the accumulators had to be taken into account.

8.3.2 Attitude Control, Communication System

Like all probes, the JWST is also three-axis stabilized. The following are used for attitude control:

- 3 sun sensors, and stars up to the 6th magnitude are tracked. These recordings are compared with a star map.
- 3 star sensors,

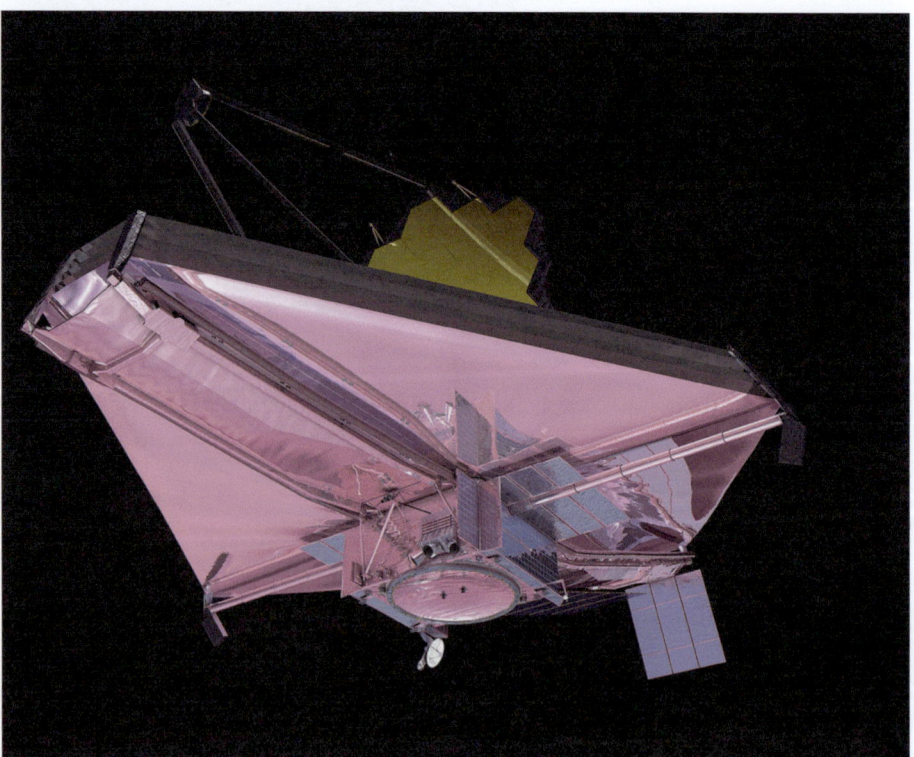

Fig. 8.9 The supply unit below the sunshield. NASA

8.3 Basic Supply and Instruments

- Gyroscopes and
- Control nozzles.

The star sensors observe an approximately 16-degree field in the sky. After the rough alignment, which achieves an accuracy of about 8", the fine alignment is carried out with a mirror in the optical beam path of the telescope.

The probe's distance and position determination is carried out with two unidirectional omnidirectional antennas. This allows its position to be controlled and, if necessary, changed from Earth. There is also a 20 cm diameter antenna for data transmission at 40 kbit/s. The actual transmission of the scientific data is carried out via a movable 60-cm parabolic antenna at a frequency of 26 GHz. The antenna must be realigned every 2 h 45 min. The data transmission rate is between 7 and 28 Mbit/s possible. In case of bad weather conditions on Earth, a lower transmission rate is chosen. In normal operation, the data is transmitted every 12 h for 4 h, which is at least 28.6 GB. The Deep Space Networks in Goldstone, Canberra, and Madrid are used for this purpose.

The onboard computer has a 58.9 GB SSD drive for storing data from the supply unit and scientific data.

8.3.3 Propulsion

The drives and the fuel are located at the supply unit and should be sufficient for at least 10 years of operation. There is good news. The trajectory to the Lagrange point L_2 was so precise that fuel could be saved, thereby extending the telescope's functional duration to 20 years. There is also a device for refueling the telescope. Perhaps a robot mission will be sent to the probe in the coming years to refill the tanks. However, such a mission is not currently planned. The drive consisted of two components:

- *SCAT, Secondary Combustion Augmented Thrusters:* served for control to the Lagrange point L_2 as well as for regular orbit correction. In each pair, one of the engines is redundant. Hydrazine N_2H_4 and dinitrogen tetroxide (N_2O_4) are used as oxidizer.
- *MRE-1: Monopropellant Rocket Engines:* there are eight of them, they serve for attitude control and compensation of the solar wind acting on the probe.

The supply unit with the onboard computer is not in the shadow but on the hot side, the temperature is about 27 degrees Celsius.

8.3.4 The Sunshield

To protect against the infrared (heat) radiation from the sun, there is a 21 × 14 m large sunshield. This thin foil consists of several layers, the outermost foil is coated with silicon. This radiates heat particularly well. This layer also gives the shield its pink color. The entire shield with the layers separated by about 40 cm was only deployed in space. This shield is exposed to extreme loads. The warmest parts of the JWST are up to 85 degrees Celsius warm, the coldest parts however at a temperature of -233 degrees Celsius. So there is about a 300 degree temperature difference! In addition, there are hits from micrometeorites present in space, which can tear holes in the foil. The foils have been specially reinforced so that longer cracks form rather than impact holes. This is hoped to ensure long functionality.

A test of the sunshield, which consists of several layers, is shown in Fig. 8.10.

8.3.5 The Optics

The diameter of the main mirror, which consists of 18 hexagonal segments, is 6.5 m. These only unfolded in space. The mirrors are made of beryllium; this has a low density, high strength, and a low coefficient of thermal expansion at the low temperatures in space. If an important optical component of a telescope expands when heated, the image quality deteriorates or the focal point changes. Each of the 18 segments has a mass of only 18 kg with an inner circle diameter of 1.3 m. The mirrors were vapor-coated with a 100 nm thick layer of gold. Gold has a particularly high reflection in the infrared. A thin layer of glass made of silicon dioxide is supposed to protect the soft gold layer from scratches and incoming particles. The mirrors can be individually aligned with motors. In the middle of each segment, there is another motor, actuator, that can adjust the curvature of the mirror.

> In optics, there is a *tolerance limit* for the accuracy with which a mirror (or a lens) must be ground. This tolerance limit can also be used to estimate how accurately the segments of the JWST must be aligned. The tolerance limit T_g for a good image is:
>
> $$T_g \sim \frac{\lambda}{10} \quad (8.1)$$
>
> thus one tenth of the wavelength λ, at which it is observed. Let's assume we observe in the near IR at a wavelength of 1000 nm, then $T_g = 100\,nm = 100 \times 10^{-9}\,m = 10^{-7}\,m = 0.0000001\,m$.

8.4 Instruments of the JWST

Fig. 8.10 Test of the sunshield. NASA/Chris Gunn

The M2 mirror (secondary mirror) is concave. The focal length of the system is 131.4 m.

> Let's estimate how large the diameter of the moon would be at the focal length of the James Webb Telescope. As a rule of thumb, the size of the moon or sun in cm is given by the focal length in m. A telescope with a 1 m focal length thus produces a 1 cm image of the sun or moon. Thus, with the James Webb Telescope, we get a 1.314 m image of the moon.

8.4 Instruments of the JWST

8.4.1 The Near Infrared Camera

NIRCam (Near Infrared Camera) is designed for observation in the range of 0.6 to 5 μm. Mercury-cadmium-telluride sensors are used as detectors, and the system is cooled to 50 K. The observation field of the two-part camera is 2.3'× 2.3'.

The system has filter wheels and, similar to the Hubble Telescope, there is a cover disc with which the light of a star can be covered and, for example, the dust ring surrounding a star in the near IR can be studied. This is therefore a *coronagraph*. In solar physics, such instruments are used to cover the sun with a cone diaphragm to study the corona even outside of a total solar eclipse in the visible range. However, observations from Earth are severely affected by scattered light caused by dust in the Earth's atmosphere.

8.4.2 MIRI

MIRI stands for *Mid Infrared Instrument*. This allows observations in the range between 5 and 28.3 μm. The camera consists of three 1024 by 1024 pixel sensors and is also used for spectroscopic analysis. The cooling goes down to 7 K. The instrument was built by the ESA with the US Jet Propulsion Laboratory, JPL, the ESA has 50% of the time for use. MIRI also has filter wheels and a coronagraph.

8.4.3 NIRSpec, FGS-NIRIS

NIRSpec stands for *Near Infrared Spectrograph*. The working range is between 0.6 to 5 μm. It can capture spectra from up to 200 objects simultaneously. FGS-NIRISS stands for Fine Guidance System/Near Infrared Imager and Slitless Spectrograph. The field of view is 2.2 by 2.2 arc minutes. The wavelength range is from 1.0 to 2.5 μm. In a special observation mode, it is hoped that this device will also provide insights into the composition of the atmospheres of exoplanets.

In Fig. 8.11, you can see the Near Infrared Spectrograph during assembly in the laboratory.

In Fig. 8.12, you can see an example of spectra from stars in the field of view. The light of the stars is broken down and based on the dark spectral lines, one can determine which elements are present there, from the shift of the lines, velocities (Doppler effect) and much more can be determined.

In Fig. 8.13, the scientific instruments, their observation range, and some research questions that can be investigated with them are listed and summarized.

8.5 First Images

In this section, we provide some examples of the performance capabilities of the James Webb Telescope.

Fig. 8.11 The infrared spectrograph of the JWST during assembly in the laboratory. NASA

Fig. 8.12 An example of star spectra with the infrared spectrograph of the JWST. NASA

8.5.1 Solar System

The exploration of the solar system is mostly done with space probes, but the constant observation of the planets with telescopes is also very important in order to study long-term changes, such as the shrinking of the Great Red Spot on Jupiter, etc.

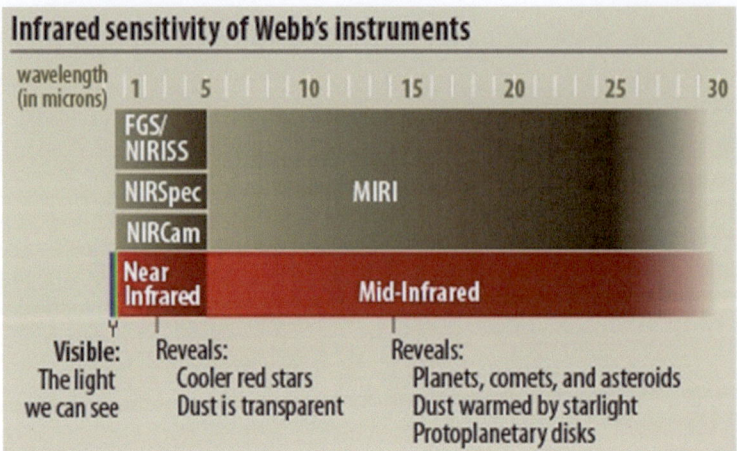

Fig. 8.13 Instruments of the JWST as well as their observation range and some research objects that can be reached with them

Figure 8.14 shows the giant planet *Jupiter* at a wavelength of 2.12 µm. On the left, you can see Jupiter's moon Europa, which casts a dark shadow next to the brightly shining Great Red Spot on the surface of Jupiter.

Figure 8.15 also shows Jupiter, taken with the James Webb Telescope; you can even see auroras as bright spots at the Jupiter poles. The image was taken using three infrared filters. The Great Red Spot appears bright in infrared light as it strongly reflects sunlight.

The Saturn moon *Titan* is the second largest moon in the solar system. It is surrounded by a dense cloud cover, but structures can be seen on the surface in the infrared. In Fig. 8.16 you can see an image of Titan. In the left image, you can see two clouds in the atmosphere. In the right image, you can see the surface through the use of three filters. The *Kraken Mare* is about the size of the Caspian Sea and consists of hydrocarbon compounds.

All gas planets have rings, but only the Saturn ring system can be observed with simple telescopes. The rings of Saturn consist of decimeter to meter-sized ice particles, which reflect sunlight very strongly and are therefore bright. The rings of the other planets consist of dust particles, which become visible in the infrared. In Fig. 8.17 you can see an image of the planet *Neptune* with the JWST.

8.5.2 Exoplanets

Exoplanets are planets outside our solar system that orbit other stars. The first exoplanets were found in the 1990s. Since exoplanets and stars orbit their common center of mass, the existence of an exoplanet around it can be inferred from a star's periodic motion. The

8.5 First Images

Fig. 8.14 Jupiter and its moon Europa. NASA, ESA, CSA, and B. Holler and J. Stansberry

Fig. 8.15 Jupiter; you can see bright spots in the polar regions, these are auroras on Jupiter. NASA/ESA

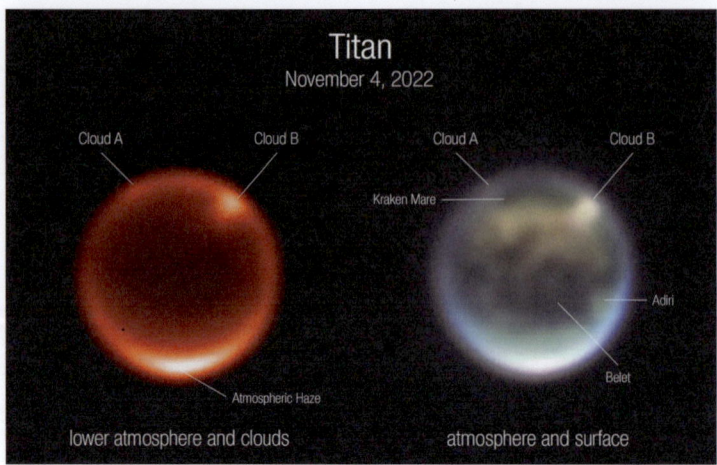

Fig. 8.16 Details in the atmosphere of Saturn's moon Titan and on its surface

Fig. 8.17 Image of Neptune with its rings that are only visible in the infrared

star's speed can be measured by the Doppler effect. Spectral lines of the star are shifted to red and blue by the periodic motion around the center of mass. This method of discovering exoplanets is also called *Radial velocity method*. If, as seen from Earth, an exoplanet passes in front of its parent star, i.e., a *Transit*, then a very small drop in the star's brightness is detected. This method of discovering exoplanets is called the transit method. There are many interconnected initiatives to find exoplanets through observations from

8.5 First Images

Earth. *WASP* stands for the Earth-bound project *Wide Angle Search for Planets*. With the help of wide-angle cameras, a large area of the sky is examined from Earth to see if there are stars whose brightness changes due to a planet transit. In Fig. 8.18, you can see the wide-angle cameras of a WASP station, the Super WASP station on *La Palma* where 8 cameras are observing the sky.

▶ The radial velocity method and the transit method are well suited for finding exoplanets.

One of the big questions of modern astrophysics is the search for *life* (see also [7]). We only know life as it occurs on Earth. However, there are strong indications that life elsewhere is at least tied to the presence of liquid water. The zone around a star where water can exist in liquid form on a planet is called the *habitable zone*. In recent years, numerous exoplanets have been found in habitable zones. To definitively determine whether there is life on a planet in the habitable zone, one should at least know the composition and physical state of its atmosphere. The problem is that due to the great distance, the exoplanets are always found in the immediate vicinity of the star and the radiation of the exoplanet is drowned out by the much brighter star. The contrast between the radiation of the hot star and the cooler exoplanet is not so extreme in the infrared. This is exactly where the James Webb Telescope comes in. In Fig. 8.19, you can see the spectrum of the exoplanet *WASP-96-b*. It is a *Transmission spectrum*. One compares:.

Fig. 8.18 Wide-angle cameras for the search for exoplanets, sometimes astrophysics can also be carried out with simple instruments

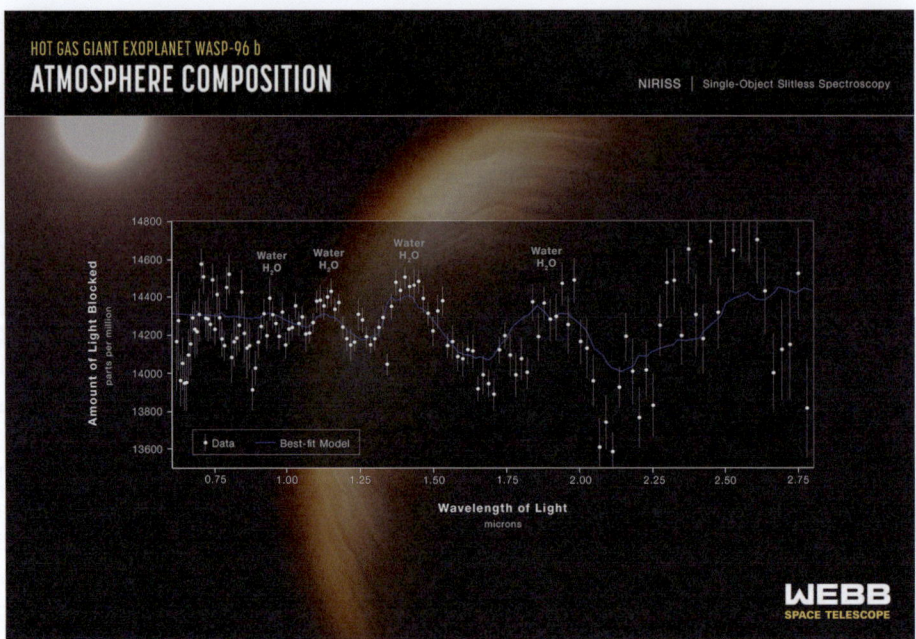

Fig. 8.19 Spectrum of an exoplanet's atmosphere. NASA, ESA, CSA, STScI

- Starlight that goes through the atmosphere of an exoplanet during a transit, where the exoplanet is in front of the star;
- Radiation of the star outside of a transit.

The depicted data points now indicate the amount of starlight that is absorbed by the planet's atmosphere. The data was obtained with NIRISS, in a wavelength range from 0.6 μm (red) to 2.8 μm (near infrared). The spectral lines of known elements and molecules can be seen. The temperature of the atmosphere follows from the amplitudes (peaks) of the lines; the hotter the atmosphere, the larger these amplitudes. Information about haze and clouds in the atmosphere follows from the shape of the lines. In this case, the temperature is about 725 degrees Celsius. WASP-96 is 1150 light years away from us, the planet WASP-96b orbits its central star at only 1/20 of the Earth-Sun distance. One orbit lasts 3.5 Earth days. The star is slightly older than the Sun but otherwise similar in mass, temperature, and size.

The light curve Fig. 8.20 shows the brightness progression during a transit. This curve was also obtained with the NIRISS instrument. Note the extremely high precision of the measurements—recognizable by the small fluctuations in brightness.

The exoplanet WASP-96-b has 1.2 times the radius of Jupiter, therefore it is 13.5 times as large as Earth. However, it only has about half the mass of Jupiter (which corresponds to 153 Earth masses).

8.5 First Images

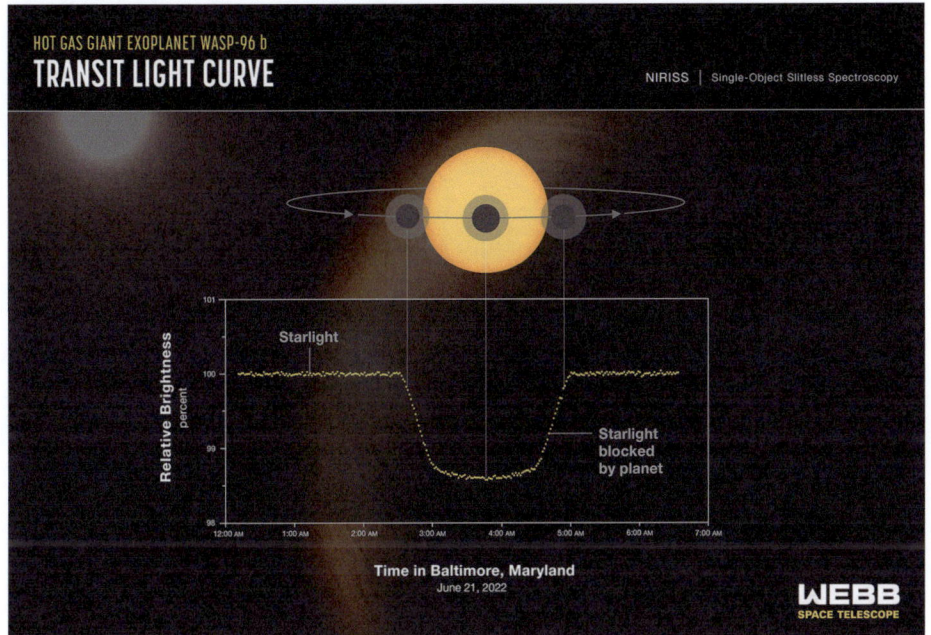

Fig. 8.20 Progression of brightness during a transit. NASA, ESA, CSA, STScI

Figure 8.21 shows another example of an exoplanet's atmospheric spectrum: WASP-39-b. Clearly signatures of water, H_2O, carbon dioxide CO_2, sulfur dioxide SO_2 and carbon monoxide CO can be seen. The planet is located in the constellation Virgo (Virgin) and is 700 light years away. It is also a planet of thetype "Hot Jupiter", these are planets with roughly the mass of Jupiter, which are very close to their parent star and thus may lose part of their atmosphere through evaporation.

8.5.3 Stars

In Fig. 8.22, you can see how the quality of the images has improved through the use of ever larger instruments. The images show the same field of sky taken with the *WISE Telescope* (Wide Infrared Survey Explorer, launched 2010; diameter of the optics 40 cm), the *Spitzer Telescope* (2003–2020, diameter 85 cm) and the *James Webb Telescope* (6.5 m).

Figure 8.23 shows the *southern Ring Nebula,* a planetary nebula. These nebulae are created by the ejection of the outer star shell at the end of the development of a star of not too great mass (like our sun). The left image shows the hot gas surrounding the central star using special filters. The right image also shows cooler gas that has expanded further into space. The left image shows the glow of hot ionized gas, the right image the

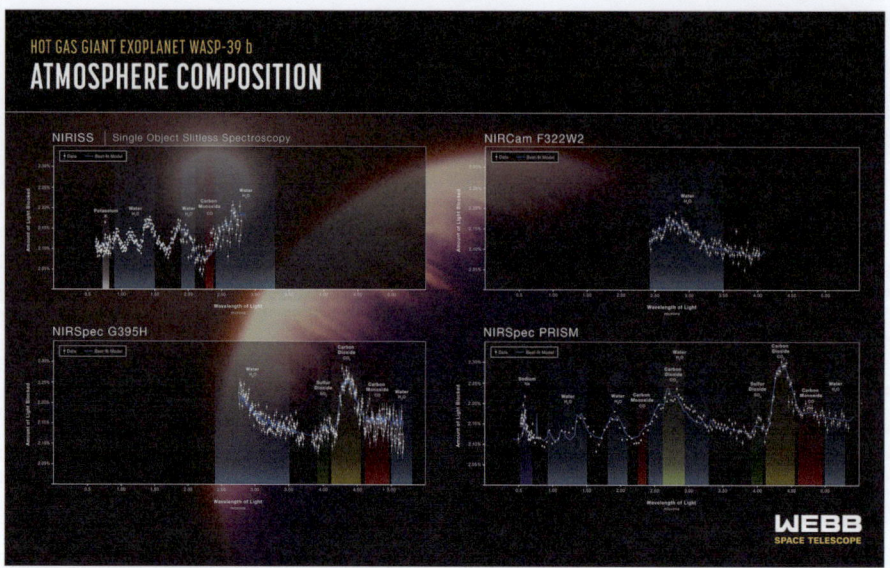

Fig. 8.21 Transmission spectrum of the exoplanet WASP-39-b. Note the signatures of water in the near infrared.. NASA, ESA, CSA, Joseph Olmsted (STScI)

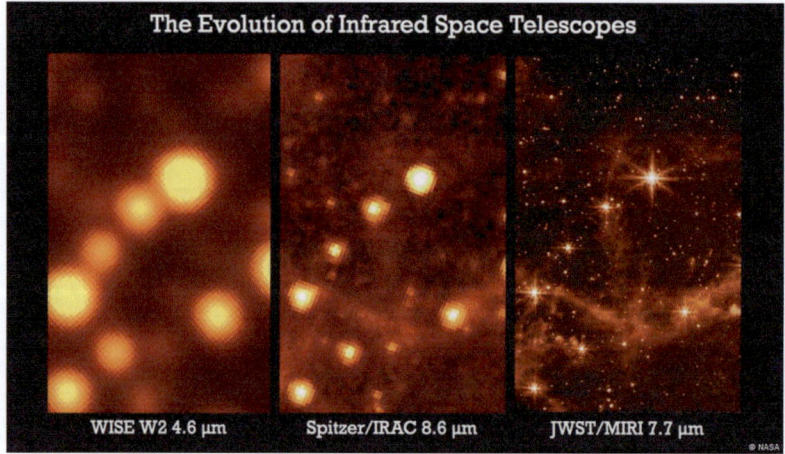

Fig. 8.22 Comparison of three images of the same sky region with different instruments. NASA

glow of cool molecules. Comparing the two images allows statements to be made about the ejection of gas masses and their expansion.

About the *Eagle Nebula,* which contains the dust clouds known as the Pillars of Creation, has already been reported on the Hubble Telescope. The James Webb Telescope

8.5 First Images

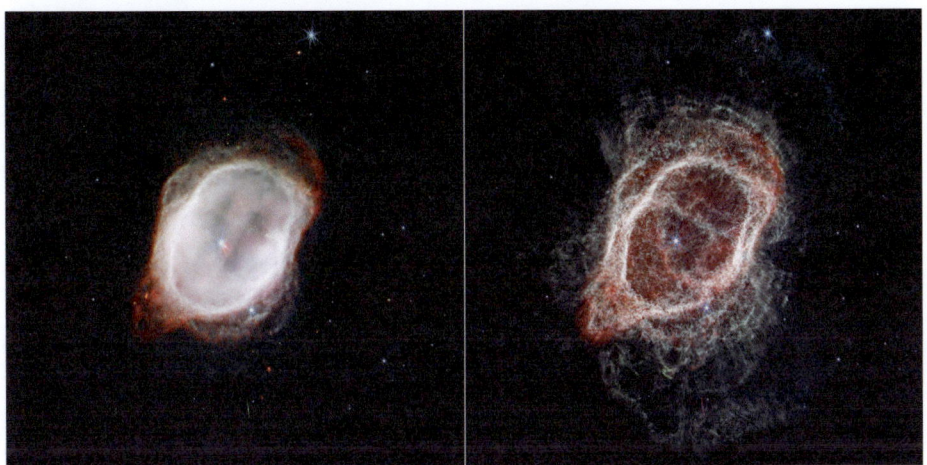

Fig. 8.23 The southern Ring Nebula. NASA, ESA, CSA, STScI, Orsola De Marco

brings a significant gain in details here (Fig. 8.24). The red structures at the tips of the pillars are light from stars hidden behind the dust clouds.

In Fig. 8.25, you can see a *protostar*, the precursor of a star, within a gas cloud. The star is surrounded by a glowing cloud from which it is still drawing mass through accretion. The star is not yet stable but is gaining mass. There are also gaps in the cloud due to matter ejections from the star. The bubble-like structure was created by sporadic ejections. The filaments consist of molecular hydrogen that was compressed by shock waves from past ejections. The cloud is located in the constellation Taurus (Taurus) at a distance of 460 light years. The size of the image corresponds to 2.2 arc minutes, which are 0.3 light years or 20,000 times the distance Earth-Sun.

8.5.4 Galaxies and the Early Universe

One of the first published images of the James Webb Telescope is given in Fig. 8.26. It shows extremely distant galaxies, as well as the effect of a gravitational lens. All objects seen in the image are galaxies, with the exception of a few stars belonging to the Milky Way. These can be easily recognized by the "spikes". These occur in the JWST just as in other mirror telescopes due to the mounting of the secondary mirror, which is attached in front of the main mirror. Point-like light sources are then distorted by the diffraction of light.

The JWST observes in the infrared. Thus, one can observe galaxies whose light is very strongly redshifted. According to Hubble's law, these galaxies are very far away from us, we are looking into the early days of the universe. In Fig. 8.27 you can see the galaxy cluster *Abell 2744*. In the enlarged image sections in the middle, you can see two

Fig. 8.24 The dust clouds in the Eagle Nebula known as the Pillars of Creation. NASA, ESA, CSA, STScI IMAGE PROCESSING: Joseph DePasquale (STScI), Alyssa Pagan (STScI), Anton M. Koekemoer (STScI)

extremely distant galaxies: In case (1) a galaxy with a redshift of $z = 10.5$, in case (2) with a redshift of $z = 12.5$. The galaxy (1) existed at a time when the universe was only 450 million years old, the galaxy (2) already existed when the universe was only 350 million years old.

This galaxy cluster has been previously studied from Earth and with other satellites. It turns out that in addition to the visible matter, there must also be a particularly large amount of invisible dark matter, which can only be detected by gravity. In Fig. 8.28, you can see a superimposition of three images. In the background is an image of the galaxy cluster taken with the Hubble Telescope, in red an image of the cluster in X-ray light, and in blue the possible distribution of dark matter (from computer simulations). The visible luminous matter only makes up about 5% of the total mass of the cluster. The

8.5 First Images

Fig. 8.25 A protostar within a gas cloud (L1527)

gas makes up about 20% of the cluster and is very hot, therefore it shines in X-ray light (red). The galaxy cluster Abell 2477 itself is about 3 billion light years away from us. Collisions between galaxies of this cluster have released stars and gas masses, the latter are responsible for the strong X-ray radiation excited by these stars. There are probably about 200 billion such ejected stars. About 350 million years ago, there was probably a collision of four smaller galaxy clusters, as computer simulations show. The cluster is sometimes also referred to as Pandora's galaxy cluster.

In Fig. 8.29, the quasar SDSS J165202.64+172852.3 is depicted. Quasars are very brightly shining cores of galaxies. If you only see the core, in whose center there is a *supermassive black hole* of several million solar masses, then you get the impression that

Fig. 8.26 First Deep Field with the James Webb Telescope. The galaxies around the center of the image appear to be distorted by the gravitational lens effect. NASA

it is a star. The great brightness of the core comes from the collapse of matter into the black hole. On Because of their brightness, quasars can be seen at very great distances. In the image, you can see on the left a picture of the sky area with the quasar from the Hubble Telescope. On the right, you can see images that were made with different filters of the JWST NIRSpec. The images show different areas of the line profile of the doubly ionized oxygen line. The red color indicates images that were made in the red-shifted area of the line. This indicates matter that is moving away from us. The blue images indicate matter that is moving towards us. Green indicates that the matter is not moving in relation to us. The quasar is about 11 billion light years away from us. Due to the high

8.5 First Images

Fig. 8.27 Extremely young galaxies. NASA, ESA, CSA, Tommaso Treu (UCLA) IMAGE PROCESSING: Zolt G. Levay (STScI)

Fig. 8.28 The galaxy cluster Abell 2477 overlaid with an X-ray image and (colored in red) and the distribution of dark matter (blue). NAS/ESA, J. Merten, D. Coe

redshift, the oxygen line is seen in the near infrared, normally it is at a wavelength of about 500 nm, i.e., in the blue-green range.

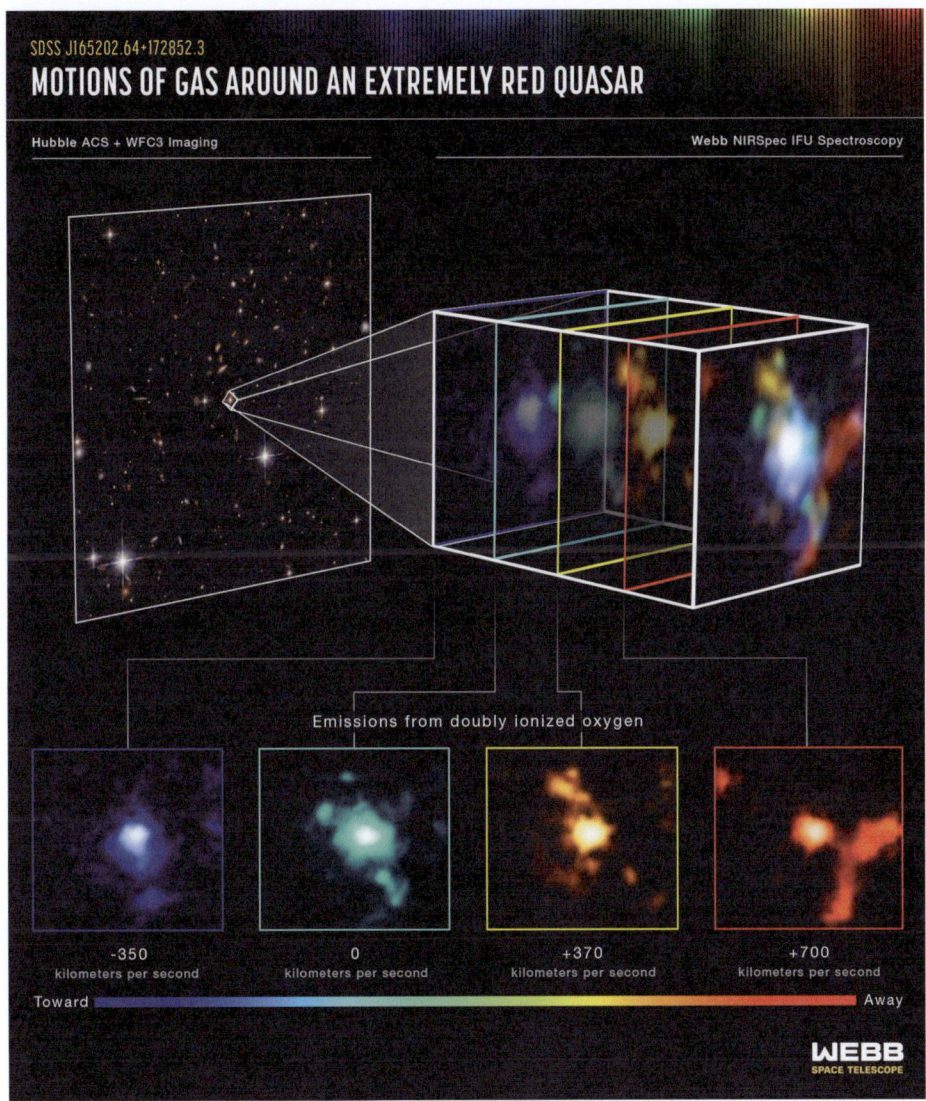

Fig. 8.29 Dynamics of the gas (ionized oxygen) around a quasar. NASA, ESA, CSA, STScI

References

1. S. Anderl. *Dunkle Materie*. C. H. Beck, 2022. Das große Rätsel der Kosmologie.
2. Jeffrey O. Bennett, M. Donahue, Nicholas Schneider, and Mark Voit. *The Cosmic Perspective: The Solar System*. 2020.
3. D. Giulini and K. Kiefer. *Gravitationswellen*. Springer, 2016. Einblicke in Theorie, Vorhersage und Entdeckung.
4. D.J. Griffiths. *Einführung in die Physik des 20. Jahrhunderts*. Pearson Studium, 2015. Relativitätstheorie, Quantenmechanik, Elementarteilchenphysik und Kosmologie.
5. Arnold Hanslmeier. *Faszination Astronomie: Ein topaktueller Einstieg für alle naturwissenschaftlich Interessierten*. 2016.
6. Arnold Hanslmeier. *Das helle und das dunkle Universum*. 2017.
7. Arnold Hanslmeier. *Planetary Habitability and Stellar Activity*, volume 3. 2018.
8. Arnold Hanslmeier. *Einführung in Astronomie und Astrophysik*. 2020.
9. Arnold Hanslmeier. *Fascination Astronomy: A cutting-edge introduction for all those interested in the natural sciences*. 2023.
10. Arnold Hanslmeier. *Introduction to Astronomy and Astrophysics*. 2023.
11. D. Lorenzen. *Hubble*. Kosmos, Franckh, 2019. 30 Jahre Hubble.
12. Chr. Spiering. *Neutrinoastronomie*. Springer, 2021. Blick in verborgene Welten.
13. A. Vamplew. *Praktische Astronomie*. Dorling Kindersley Verlag, 2021. Den Sternenhimmel entdecken.

MIX
Papier aus verantwortungsvollen Quellen
Paper from responsible sources
FSC® C105338

If you have any concerns about our products,
you can contact us on
ProductSafety@springernature.com

In case Publisher is established outside the EU,
the EU authorized representative is:
**Springer Nature Customer Service Center GmbH
Europaplatz 3, 69115 Heidelberg, Germany**

Printed by Libri Plureos GmbH
in Hamburg, Germany